TEXAS TECH UNIVERSITY PRESS

This book is typeset in Times and Helvetica. The paper used in this book meets the minimum requirements of ANSI/NISO Z39.48-1992 (R1997). ∞

Printed in the United States of America

Library of Congress Cataloging-in-Publication Data
Jones, J. Knox.
    Illustrated key to skulls of genera of North American land mammals /
J. Knox Jones, Jr. and Richard W. Manning.
        p.   cm.
    Includes bibliographical references.
    ISBN 0-89672-289-9 (pbk.)
    Mammals—North America—Anatomy—Classification. 2. Skull—
Anatomy—Classification. I. Manning, Richard W. II. Title.
QL739.J66 1992
599'.0471—dc20   92-18804
                                                                          CIP

Texas Tech University Press
Box 41037
Lubbock, Texas 79409-1037 USA
800.832.4042
ttup@ttu.edu
www.ttup.ttu.edu

CONTENTS

ADDENDA

Since the first printing, the following taxonomic and nomenclatorial changes have occured:

Plecotus: New World species of this taxa currently are referred to the genus *Corynorhinus* and are considered distinct from Old World species of *Plecotus*. (R. Tumlison and M. E. Douglas. 1992. Parsimony analysis and the phylogeny of the plecotine bats (Chiroptera: Vespertilionidae). J. Mamm., 73:276-285.)

Bison: This genus currently is not recognized as distinct from *Bos*. (J. K. Jones, Jr., R. S. Hoffmann, D. W. Rice, C. Jones, R. J. Baker, and M. D. Engstrom. 1992. Revised checklist of North American mammals north of Mexico, 1991. Occas. Papers Mus., Texas Tech Univ., 146:1-23.)

INTRODUCTION

Skulls and teeth of North American mammalian genera have been described in numerous publications, many of which contain keys to various of the taxa. However, few of these incorporate illustrations designed to acquaint the inquiring student with pertinent characteristics by which genera can be recognized readily, or in some cases otherwise appropriate references are too bulky to use with ease in the laboratory, or do not focus on identifications at the generic level. For all of these reasons and more, we decided, in the course of teaching laboratories in mammalogy, to devise an illustrated key to skulls of those North American mammalian genera not belonging exclusively to marine groups in the hope it would provide a better laboratory experience for students than currently is available to many of them.

Our treatise includes all presently recognized genera of native terrestrial mammals occurring in North America to the north of México, although it generally is applicable to much of the temperate part of that country as well. We have excluded marine groups—the Cetacea, Sirenia, and the families of "pinniped" carnivores—because most courses in mammalogy do not deal extensively with marine mammals. We thus leave it to those instructors who do so to supplement the materials presented here as may be necessary to their own specific needs.

Following the key to included orders, each order is introduced with some basic information as to distribution, diversity, and general characteristics of the group, followed by a key to families except for those two orders (Didelphimorphia and Xenartha) that contain but one family in the geographic area of coverage. Similarly, each family is briefly introduced, followed by, as necessary, a key to its included genera. Information on orders and families is presented to strengthen, by way of repetition, information presented in lectures or in the text for the course. Additional material on anatomical characteristics and ecological diversification at the ordinal and familial levels can be found in such sources as Anderson and Jones (1984), DeBlase and Martin (1980), and Vaughan (1986). Nowak (1991) is a useful reference focused on mammalian diversity at the generic level, as is Hall (1981) for general coverage of all North American mammals.

Recognizing that there still is controversy concerning the generic as opposed to subgeneric status of some lineages—the chipmunks, *Tamias* and *Eutamias,* or among felids, for example—we have alluded in some such instances to characteristics that can be used to distinguish taxa should an instructor wish to do so. Individual species are mentioned only occasionally. To construct a key to North American species would,

1

in some instances, require inclusion of external as well as cranial features, especially if multiple characteristics are used; this can be done rather easily for local faunas should such information be desirable for students in a particular class setting. Nor have we included domesticated mammals or free-ranging exotics, with the exception of several widespread and common rodent genera introduced to North America from other continents. Insofar as morphology of skulls and teeth is concerned, we avoided mention of some otherwise useful characters, well known to specialists, simply because we felt their inclusion went beyond expectations that the average undergraduate could use them with confidence.

In constructing the keys, we tried, wherever possible, to include two or more pairs of contrasting characters for two reasons: first to encourage students to examine features of mammalian skulls more completely than would be the case if only single character states were used, and second to accommodate identification of specimens for which skulls may be incomplete or may have missing teeth. Use of multiple characters does have at least one serious drawback, however, in that it occasionally precludes use of one character that, by itself, clearly contrasts between two groupings, because of an earlier split in the key.

Keys are dichotomous and necessarily pertain to adult specimens, although most characters used also apply to young individuals. Each key consists of pairs of contrasting statements, which are referred to as couplets. Each couplet is numbered to the left with the same numeral, one bearing a prime sign. Beginning at the top of each key, characteristics of a specimen under study should best fit those listed in one or the other of the two units of each couplet. At the right margin, is the number of the couplet in the key to which the user is directed next or the name of the taxon to which the specimen belongs. By following carefully through progressive couplets, it should be possible to identify a skull in hand with relative ease. By their very nature, however, keys can be, and frequently are, a matter of contention with respect to use of one or another set of contrasting characters. The authors invite criticisms of this sort, which will be duly considered when a new edition of this booklet is prepared.

Technical terms used in the keys and accompanying text are defined in a glossary, to which students should refer regularly. Several figures in the glossary will aid in refreshing the memory of users as to location or appearance of certain cranial characters. A hand lens or low-power binocular microscope will be needed to adequately appreciate some features, as will a pair of calipers or a millimeter ruler for use when measurements are mentioned (all measurements listed in keys are in millimeters). Dental formulae are used at some places in keys or individual teeth, such as the first upper incisor or first lower molar, are mentioned by name. The statement "incisors 2/1," for example, means there are two incisors in each half of the upper jaw and one in each half of the lower jaw, as explained in the glossary; "cheekteeth 5/4" means there are five cheekteeth (premolars and molars) on each side above and four on each side below.

Another way to refer to individual teeth, which we have tried to avoid but which occasionally appears parenthetically for clarification, is to allude to them by pristine type and number based on the primitive placental dental formula of three incisors, one canine, four premolars, and three molars on each side both above and below, using capital letters to denote upper teeth and lower case letters to identify those in the mandible (see Fig. 49 in the glossary). No North American placental land mammals have the pristine number of 44 teeth with the exception of three genera of moles (Table 1). Loss of teeth in the evolution of mammalian groups generally has taken place from back to front in incisors (although variable) and molars, and from front to back in premolars. Thus, the two remaining premolars in the upper jaw of most sciurid rodents are P3 and P4, whereas the one remaining molar tooth above and below in cats is the M1 and m1, respectively.

We have not illustrated all included genera. Illustrations, both photographs and line drawings (some adapted from other published sources), were selected to illustrate certain special or contrasting features of skulls, especially where these are not likely to be easily appreciated by the uninitiated student. Some illustrations are intended primarily to illustrate the diversity in size and shape of crania among related genera. Four tables of dental formulae also are included.

Instructors should be able to use illustrations to emphasize key characters as well as important features not used in keys, and the perceptive student will quickly note certain differences between or among taxa that are not otherwise covered in the text. In Figure 25, for example, the differences in the basioccipital regions and auditory canals used in the key to rodent families to distinguish between Castoridae (*Castor*) and Erethizontidae (*Erethizon*) are easily appreciated, but note also the relative size of the incisors and auditory bullae, and the marked differences between the palatal and mesopterygoid regions, among other differential characters. For this and related reasons, ample margins are provided in the keys for students to add such notes as may be useful to them.

Photographs were taken by the second author. We are grateful to N. L. Olson for his assistance in printing them. We used specimens housed in the Museum of Texas Tech University and the Museum of Natural

History at the University of Kansas; R. M. Timm kindly arranged for access to, or loan of, those from the latter institution. For reading and critically commenting on a draft version of the manuscript, we especially acknowledge our colleagues Robert C. Dowler, Robert E. Martin, Frederick B. Stangl, Jr., and Kenneth T. Wilkins.

At Texas Tech University, Clyde Hendrick, dean of the Graduate School, assisted us by providing (to Manning) a research assistantship in the summer of 1989, Gary Edson, director of the Museum, supplied logistical support, and Shirley Burgeson skillfully typed the many versions of the manuscript. To all those named, and to others who assisted our efforts in one way or another, we extend our appreciation.

KEY TO ORDERS OF NORTH AMERICAN LAND MAMMALS

1. Incisors 1/1 or 2/1 (second upper incisor peglike and set directly behind the first); broad diastema between incisors and cheekteeth; first upper incisor enlarged, evergrowing, modified for gnawing 2

1'. Incisors variable (never 1/1, 2/1 only in one insectivore and then second upper incisor set normally in toothrow) to absent; if incisors present both above and below, no diastema, or only a modest diastema, in toothrow; first upper incisor, if present, usually not noticeably enlarged, not evergrowing, not modified for gnawing 3

2. Incisors 2/1; maxillary fenestration(s) present . . . . **Lagomorpha**

2'. Incisors 1/1; no maxillary fenestrations . . **Rodentia**

3. Incisors lacking; cheekteeth peglike, lacking enamel . . . . **Xenarthra**

3'. Incisors present in at least lower jaw; cheekteeth not peglike, composed partly of enamel 4

4. Incisors 5/4; angular process of dentary distinctly inflected . . . . **Didelphimorphia**

4'. Incisors never more that 3/3; angular process of dentary not inflected 5

5. Upper incisors absent (except Dicotylidae); upper canine absent or, if present, separated from cheekteeth by modest diastema; postorbital bar present (incomplete in Dicotylidae) . . . . . **Artiodactyla**

5'. Upper incisors present; upper canine present and not separated from cheekteeth by diastema; postorbital bar absent 6

6. Molars 3/3, except in two bats (*Diphylla* and *Lep-tonycteris*) in which they are 2/2, last upper molar never dumbell-shaped; number of incisors variable; greatest length of skull less than 45 mm (usually much less) 7

6′. Molars never 3/3 (1/1, 1/2, 2/2, or 2/3); upper incisors always three in number; greatest length of skull usually more than 45 mm (if less, then last upper molar dumbell-shaped) . . . . **Carnivora**

7. Canine poorly to modestly developed, never largest tooth in anterior part of toothrow except in *Condylura* (in which there are 44 total teeth); dorsal profile of skull flattened, braincase not much higher than rostrum . . . . **Insectivora**

7′. Canine well developed, largest tooth in anterior part of toothrow except in *Diphylla* (in which bladelike canine is slightly smaller than the similarly shaped first incisor); total number of teeth never more than 38; dorsal profile of skull not flattened, braincase noticeably higher than rostrum except in a few molossids . . . . **Chiroptera**

Order DIDELPHIMORPHIA
Opossums

Marsupials are known only from North and South America (where a single species, *Didelphis virginiana*, is found to the north of México) and from the Australian region. Aside from the egg-laying monotremes, they are the most primitive of living mammals. The fossil history of the group dates back certainly to the late Cretaceous of the New World.

Most recent authors claim at least supraordinal status for Marsupialia, which is reasonable considering the tremendous fossil and Recent diversity among these mammals. We employ Didelphimorphia as an ordinal name following Marshall *et al*. (1990), who summarized the history of classification of the group and the conjecture surrounding it. Modern marsupials encompass some 75 genera and approximately 250 species. There are three living families in the New World, but only one, Didelphidae, occurs in North America.

Among living groups, marsupials or metatherians differ from eutherians in a variety of characteristics. Those most germane to the present work are: braincase narrow and relatively small; angular process of dentary distinctly inflected; jugal large and extending posteriorly to form preglenoid crest; posterior border of palate deflected downward, frequently ridged; cheekteeth primitively of three premolars and four molars (the reverse is true in placental mammals); incisors never of the same number above and below (in *Didelphis* the incisor formula is 5/4, whereas it never is more than 3/3 in placentals). *Didelphis* has a total of 50 permanent teeth. See Figs. 1 and 2.

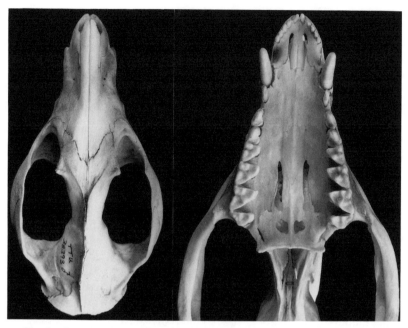

Fig. 1.—Dorsal and ventral views of cranium of *Didelphis* (greatest length, 110.3 mm).

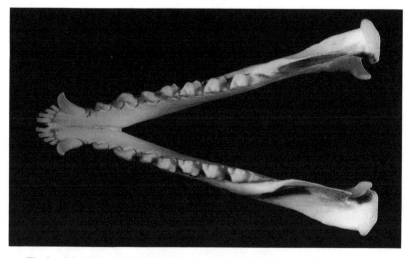

Fig. 2.—Mandible of *Didelphis*. Note teeth and inflected angular processes.

Order INSECTIVORA
Insectivores

This order includes the most primitive of living placental mammals, although some, such as moles, are highly specialized, and represents the general stock from which more advanced placental groups are thought to have evolved. The fossil record dates from the late Cretaceous, about 75 million years ago.

Modern insectivores comprise six families, 61 genera, and some 375 species. Representatives occur on all continents except Antarctica and Australia, although in South America they are limited to the northwestern part of the region, south to Peru. Two families, Soricidae and Talpidae, occur in the New World, the former represented north of México by four genera and the latter by five.

Key to Families of Insectivora

1. Teeth pigmented; zygomatic arch absent; tympana ringlike, lacking tympanic bulla . . . . . **Soricidae**

1'. Teeth unpigmented; zygomatic arch present; tympanic bulla present, complete or incomplete
. . . . **Talpidae**

Soricidae (shrews)

Shrews have the same broad distribution as given above for the order Insectivora. Of the four genera occurring in temperate North America, *Cryptotis* and *Notiosorex* are represented there by a single species each, *Blarina* encompasses three closely related species, and about 25 species are relegated to the genus *Sorex*, which occurs also in the Old World. Of the latter, one (*Sorex hoyi*) was until rather recently regarded as the sole representative of *Microsorex*, now recognized as a subgenus.

The teeth of shrews between the enlarged first incisor and the large fourth premolar are of uncertain origin. All are relatively simple teeth that are referred to as unicuspids, and their number and size are of taxonomic importance at both the generic and specific levels. The deciduous precursors as

there may be of these teeth are shed before birth. The total
number of permanent teeth in North American soricids
varies from 28 to 32.

Key to Genera of Soricidae

1. Five unicuspid teeth in upper jaw, three to five visible
   in lateral view   2
1'. Less than five unicuspid teeth in upper jaw, all usually
   visible in lateral view   3

2. Four unicuspids visible in lateral view; lateral border
   of braincase forming a sharp, pointed angle
   . . . . *Blarina*
2'. Three (subgenus *Microsorex*), rarely four, or usually
   five unicuspids visible in lateral view; lateral border
   of braincase smoothly rounded   . . . . *Sorex*

3. Four unicuspid teeth in upper jaw; teeth relatively
   heavily pigmented   . . . . *Cryptotis*
3'. Three unicuspid teeth in upper jaw; teeth only lightly
   pigmented   . . . . *Notiosorex*

Talpidae (moles)

Talpids are found throughout much of temperate North
America except in the montane west, south to northernmost
México, across the temperate parts of the Palearctic from
England to Japan, and southward into the Oriental Realm to
the Malay Peninsula. All five genera that occur in the New
World are found exclusively there. Four of these (*Condylura,
Neurotrichus, Parascalops,* and *Scalopus*) are monotypic.
The fifth, *Scapanus* of the Pacific Coast, contains three
species. Teeth in the five genera of moles total 36 or 44 (see
Table 1).

Table 1. Dental formulae of five genera of moles.

Genus ·	Incisors	Canines	Premolars	Molars	Total
Neurotrichus	2/1	1/1	3/4	3/3	36
Scalopus	3/2	1/0	3/3	3/3	36
Condylura	3/3	1/1	4/4	3/3	44
Parascalops	3/3	1/1	4/4	3/3	44
Scapanus	3/3	1/1	4/4	3/3	44

Key to Genera of Talpidae

1. Upper incisors distinctly procumbent, smaller than canines . . . . *Condylura*

1′. Upper incisors only slightly or not at all procumbent, first incisor largest tooth in anterior part of toothrow 2

2. Four or five smaller teeth in upper jaw between first incisor and last premolar; total of 36 teeth 3

2′. Six smaller teeth in upper jaw between first incisor and last premolar; total of 44 teeth 4

3. Tympanic bulla incomplete; first premolar larger than canine; greatest length of skull less than 25 mm . . . . *Neurotrichus*

3′. Tympanic bulla complete; first premolar smaller than canine; greatest length of skull more than 28 mm . . . . *Scalopus*

4. Tympanic bulla incomplete; first upper incisor with posterior accessory cusp; canine larger than adjacent teeth . . . . *Parascalops*

4′. Tympanic bulla complete; no accessory cusp on first incisor; canine about same size as adjacent third incisor . . . . *Scapanus*

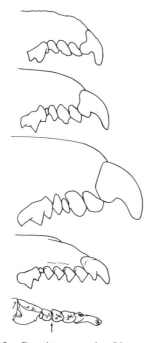 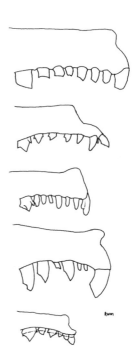

Fig. 3.—Drawings to scale of lateral views of anterior toothrows (first incisor to last premolar) of shrews. From top to bottom: *Notiosorex, Cryptotis, Blarina, Sorex,* and *Sorex (Microsorex),* which is in occlusal view to show the minute third (arrow) and fifth unicuspids. Partially modified from Jones *et al.* (1983).

Fig. 4.—Drawings to scale of lateral views of anterior toothrows (first incisor to last premolar) of moles. From top to bottom: *Scapanus, Condylura, Parascalops, Scalopus,* and *Neurotrichus.*

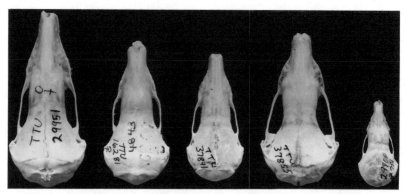

Fig. 5.—Dorsal views of crania of moles. From left to right: *Scapanus, Condylura, Parascalops, Scalopus,* and *Neurotrichus.* The three genera to the left have 44 total teeth, the two on the right, 36. For reference, greatest length of skull of the *Scapanus* is 43.5 mm, that of the *Neurotrichus* is 21.9 mm.

Order CHIROPTERA
Bats

Bats are the only mammals that are truly volant, setting them easily apart from all other groups by virtue of characteristics associated with flight. They occur on all continents save Antarctica, and on many islands, some of which (Hawaii and New Zealand, for example) are rather far removed from continental land masses. Fossil remains of bats have been found as far back as the early Eocene, some 50 million years ago. Modern chiropterans are divided into two suborders, the Megachiroptera of the warmer parts of the Old World and Microchiroptera, which has essentially the same geographic distribution as the order. There are 17 families and about 170 genera and 900 species.

Four families of bats occur in North America to the north of México: Mormoopidae (one species in the area of coverage), Phyllostomidae (five), Vespertilionidae (30), and Molossidae (six). See Table 2 for dental formulae of genera included in this work.

Key to Families of Chiroptera

1. Braincase broader than long; skull distinctly concave in lateral view, rostrum noticeably depressed basally
. . . . **Mormoopidae**

1'. Braincase longer than broad; skull not concave in lateral view, rostrum not depressed 2

2. Premaxillae fused anteriorly, no anterior palatal emargination; paired incisive foramina present; incisors 2/2 or 2/0 . . . . **Phyllostomidae**

2'. Premaxillae separated anteriorly or barely in contact, palate emarginated (except *Eumops*); paired incisive foramina absent; incisors 2/3, 1/3, or 1/2 3

3. Palate broadly emarginated anteriorly; palate
    noticeably extended posteriorly, terminating well
    behind third upper molars; premolars 1/2, 1/3, 2/2
    (*Euderma, Lasiurus, Pipistrellus*), or 2/3; incisors
    1/2 (*Antrozous*), 1/3 (*Lasiurus, Nycticeius*), or 2/3
    . . . . **Vespertilionidae**

3′. Palate only modestly, narrowly, or not at all emar-
    ginated; palate not noticeably extended posteriorly,
    terminating only slightly behind third upper molars;
    premolars 2/2; incisors 1/2 or 1/3   . . .   **Molossidae**

Table 2. Dental formulae of 18 genera of bats.

Family and genus	Incisors	Canines	Premolars	Molars	Total
Family Mormoopidae					
Mormoops	2/2	1/1	2/3	3/3	34
Family Phyllostomidae					
Diphylla	2/2	1/1	1/2	2/2	26
Choeronycteris	2/0	1/1	2/3	3/3	30
Leptonycteris	2/2	1/1	2/3	2/2	30
Macrotus	2/2	1/1	2/3	3/3	34
Family Vespertilionidae					
Antrozous	1/2	1/1	1/2	3/3	28
Nycticeius	1/3	1/1	1/2	3/3	30
Lasiurus	1/3	1/1	1/2 or 2/2	3/3	30-32
Eptesicus	2/3	1/1	1/2	3/3	32
Euderma	2/3	1/1	2/2	3/3	34
Pipistrellus	2/3	1/1	2/2	3/3	34
Idionycteris	2/3	1/1	2/3	3/3	36
Lasionycteris	2/3	1/1	2/3	3/3	36
Plecotus	2/3	1/1	2/3	3/3	36
Myotis	2/3	1/1	2/3 or 3/3	3/3	36-38
Family Molossidae					
Eumops	1/2	1/1	2/2	3/3	30
Nyctinomops	1/2	1/1	2/2	3/3	30
Tadarida	1/3	1/1	2/2	3/3	32

Mormoopidae (ghost-faced bat and allies)

Bats of this family occur in much of tropical America, including the Antilles, and are found northward to Texas and Arizona. Of the two recognized genera, only one genus and species, *Mormoops megalophylla*, reaches the United States. See Fig. 6.

Phyllostomidae (leaf-nosed and vampire bats)

Phyllostomids are endemic to the New World tropics and adjacent regions. The family is a morphologically diverse group, containing six subfamilies and about 45 genera. Four genera of three subfamilies reach the southwestern United States, three genera only in the extreme Southwest. A fifth genus, the vampire *Desmodus*, is known from sub-Recent remains from Texas.

Key to Genera of Phyllostomidae

1. Rostrum noticeably shortened; skull compact, relatively broad and high; first upper incisor and canine enlarged, bladelike . . . . *Diphylla*

1'. Rostrum not shortened, of normal length or elongated; skull not compact or noticeably broad; first upper incisor and canine normal, not enlarged and bladelike 2

2. Postcanine teeth in both upper and lower jaws reduced in size, laterally compressed, evident gaps between premolars; rostrum noticeably elongated 3

2'. Postcanine teeth not laterally compressed or reduced in size, no evident gaps between premolars; rostrum not elongated . . . . *Macrotus*

3. Rostrum about same length as braincase; zygomatic arches present; incisors 2/2 . . . . . *Leptonycteris*

3'. Rostrum longer than braincase; zygomatic arches absent; incisors 2/0 . . . . *Choeronycteris*

Vespertilionidae (common bats)

Vespertilionids have the broadest distribution of any family of bats, occupying, with the exception of some islands, the same range as that of the order Chiroptera. There are some 34 genera, 10 of which are found in the New World to the north of México. In the following key, *Myotis* falls out at two different places because of intrageneric variation in the dental formula.

Key to Genera of Vespertilionidae

1. Incisors 1/2; total number of teeth 28   . . ***Antrozous***
1'. Incisors 1/3 or 2/3; total number of teeth 30 to 38   . 2

2. Incisors 1/3; total number of teeth 30 or 32   3
2'. Incisors 2/3; total number of teeth 32 to 38   4

3. Rostrum shorter than braincase, noticeably inflated anteriorly; pterygoids not parallel (broader posteriorly); upper incisor in contact with canine
     . . . . ***Lasiurus***
3'. Rostrum about same length as braincase, not inflated anteriorly; pterygoids parallel; upper incisor not in contact with canine   . . . . ***Nycticeius***

4. Premolars 1/2; total of 32 teeth   . . . . . ***Eptesicus***
4'. Premolars 2/2, 2/3, or 3/3; total of 34 to 38 teeth . . 5

5. Premolars 2/2; total of 34 teeth   6
5'. Premolars 2/3 or 3/3; total of 36 to 38 teeth   7

6. Greatest length of skull more than 16 mm; zygoma dorsally expanded at about midpoint, forming distinct flange; first upper incisor simple   . . . . . ***Euderma***
6'. Greatest length of skull less than 14 mm; zygoma thin and straight; first upper incisor with well-developed secondary cusp . . . . ***Pipistrellus***

7. Premolars 3/3; total of 38 teeth   . . . . ***Myotis***

7'. Premolars 2/3; total of 36 teeth 8

8. Auditory bulla noticeably inflated, about same size as foramen magnum; zygomatic arches equal to, or narrower than, braincase as viewed from above . . . 9

8'. Auditory bulla not noticeably inflated, smaller than foramen magnum; zygomatic arches broader than braincase as viewed from above 10

9. Breadth of braincase more than half greatest length of skull; rostrum with pronounced lateral projection directly above infraorbital foramen . . *Idionycteris*

9'. Breadth of braincase less than half greatest length of skull; rostrum relatively smooth laterally . . *Plecotus*

10. Rostrum broadened and inflated (in comparison to *Myotis*), upper surface with distinct paired concavities between lacrimal region and nares; small but distinct flange on zygoma near midpoint
. . . . *Lasionycteris*

10'. Rostrum not broadened or inflated, upper surface lacking concavities; no flange on zygoma . . *Myotis*

Molossidae (free-tailed bats)

Bats of this family, easily recognized by the "free tail" that extends far beyond the posterior border of the uropatagium, occur throughout much of the range of the order. They are not, however, distributed as far northward in the Holarctic as are vespertilionids. There are 12 modern genera and about 85 species. Three genera, *Eumops* (three species), *Nyctinomops* (two), and *Tadarida* (one) are found in North American north of México.

Key to Genera of Molossidae

1. Incisors 1/3; rostrum relatively broad and compact, not tubelike or distinctly narrowed posteriorly
. . . . *Tadarida*

1'. Incisors 1/2; rostrum relatively narrow, especially posteriorly, and tubelike 2

2. Skull relatively broad and heavy, with strongly developed lambdoidal crest; first upper premolar small, crowded to labial edge of toothrow or at least displaced labially (*E. perotis*); zygomatic breadth more than 13.5 mm . . . . ***Eumops***

2′. Skull relatively narrow and delicate, with weakly developed lambdoidal crest; first upper premolar of normal size and in line with other teeth; zygomatic breadth less than 13.5 mm . . . . . . ***Nyctinomops***

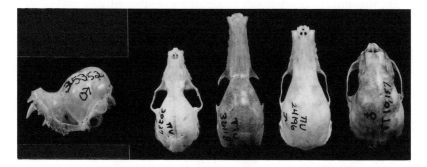

Fig. 6.—To the left is a lateral view of the cranium of *Mormoops* (greatest length, 15.5 mm). To the right are dorsal views to scale of crania of phyllostomid bats (*Macrotus, Choeronycteris, Leptonycteris,* and *Diphylla,* left to right). Greatest length of skull of the *Choeronycteris* is 30.6 mm.

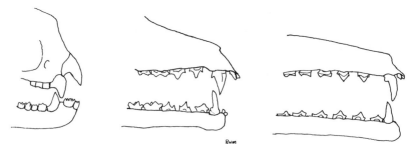

Fig. 7.—Drawings to scale of anterior parts of skulls, with toothrows, of *Diphylla, Leptonycteris,* and *Choeronycteris,* left to right.

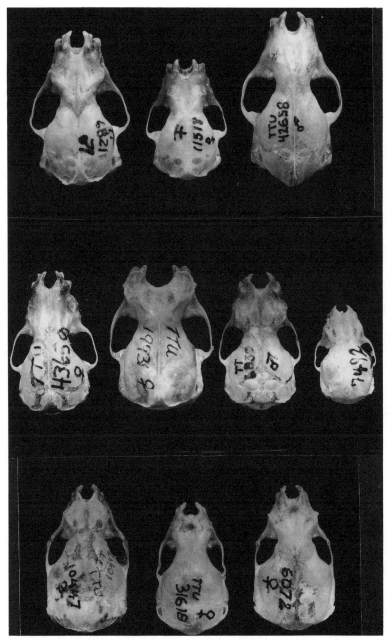

Fig. 8.—Dorsal views to scale of crania of vespertilionid bats. Top: *Eptesicus*, *Nycticeius*, and *Antrozous*, left to right. Middle: *Myotis*, *Lasiurus*, *Lasionycteris*, and *Pipistrellus*, left to right. Bottom: *Euderma*, *Plecotus*, and *Idionycteris*, left to right. Greatest length of skull of the *Eptesicus* is 19.1 mm.

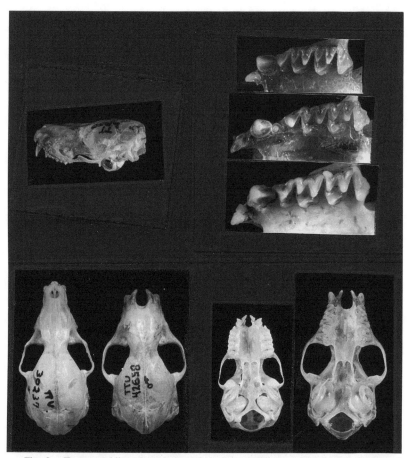

Fig. 9.—Top row: oblique-lateral view of cranium of *Lasionycteris* (left) showing rostral concavity and flange on zygoma, and occlusal views (right) to scale of left upper toothrows of (top to bottom) *Nycticeius, Myotis (velifer),* and *Eptesicus.* Bottom row: dorsal views (left) to scale of crania of *Macrotus* and *Antrozous,* and ventral views (right) to scale of crania of *Tadarida* and *Antrozous* (note terminal level of posterior extension of palate).

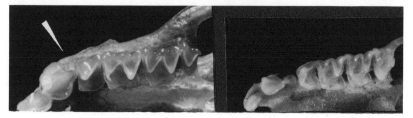

Fig. 10.—Left maxillary toothrows to scale of *Eumops* (*underwoodi*) on left and *Nyctinomops* (*macrotis*). Note small, crowded premolar (arrow) in *Eumops.*

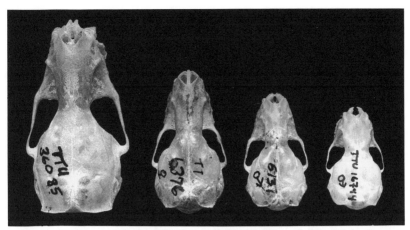

Fig. 11.—Dorsal views to scale of crania of molossid bats (left to right): *Eumops*, *Nyctinomops (macrotis)*, *Nyctinomops (femorosaccus)*, and *Tadarida*. Greatest length of skull of the *Eumops* is 31.5 mm.

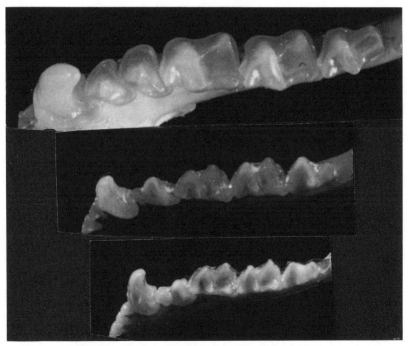

Fig. 12.—Right mandibular toothrows to scale of (top to bottom) *Eumops*, *Antrozous*, and *Lasiurus* showing variation in number, size, and configuration of teeth.

Order XENARTHRA
Xenarthrans

Xenarthra, formerly known under the ordinal name Edentata, is a New World group that dates back to the early Tertiary (Paleocene) of South America. There are four Recent families, only one of which, Dasypodidae, reaches temperate North America. It is represented there by a single species, *Dasypus novemcinctus*, the nine-banded armadillo. Overall, xenarthrans encompass 13 Recent genera and about 30 Recent species.

Obvious cranial characters by which *D. novemcinctus* can be distinguished from other mammals occurring to the north of México are: incisors and canines absent; premolars and molars subcylindrical (peglike), lacking enamel, single-rooted, seven or eight in each quadrant; rostrum elongated, tubelike; dentary narrow and elongated; tympanic bulla ringlike, incomplete. See Fig. 13.

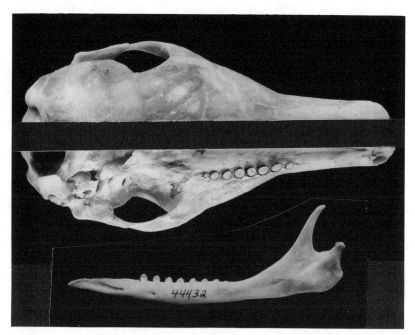

Fig. 13.—Dorsal view of left half of armadillo (*Dasypus*) cranium (top), ventral view of left half (middle), and lateral view of right dentary (bottom). Greatest length of skull is 95.5 mm.

Order LAGOMORPHA
Lagomorphs

Members of this order occur on all continents except Australia (where they have been introduced) and Antarctica. Lagomorphs have a fossil history that dates back to the Paleocene, when they evidently diverged from early rodentlike mammals. There are two modern families—Ochotonidae (one genus, 19 species) and Leporidae (11 genera, about 45 species)—both of which occur in (but neither is restricted to) temperate North America.

Like rodents, lagomorphs have large, ever-growing incisors and a distinct diastema between the incisors and the cheekteeth. Rodents have but one pair of upper incisors, however, whereas lagomorphs have two, the second peglike pair located directly behind the first. Another obvious difference between the two groups is that lagomorphs have fenestrations, or in the case of the ochotonids a single rostral fenestration, in the maxillary bone (and in some other cranial bones in leporids). They also have high-crowned (hypsodont) cheekteeth, a condition shared with some rodents.

Key to Families of Lagomorpha

1. Maxilla with one large fenestration; jugal long, projecting posteriorly nearly to level of external auditory meatus; supraorbitral process absent
. . . . **Ochotonidae**
1'. Maxilla with latticelike fenestrations; jugal short, projecting only slightly beyond zygomatic process of squamosal; supraorbital process present
. . . . **Leporidae**

Ochotonidae (pikas)

Pikas occur over much of the eastern two-thirds of the Palearctic Region and also in mountainous western North America, where two species of *Ochotona* are known. See Fig. 14.

Leporidae (hares and rabbits)

Hares and rabbits have the same distribution as does the order Lagomorpha. Three genera reside in North America north of México, two of which, *Brachylagus* and *Sylvilagus*, are limited to the New World. The former is monotypic, the latter is represented by seven species north of México. The widespread genus *Lepus* also contains seven species that occur in temperate North America.

Key to Genera of Leporidae

1. Anterior surface of first upper and lower premolars with one enamel re-entrant angle; auditory bulla relatively large, width equal to, or greater than, length of maxillary toothrow; palatal bridge less than 4.5 mm in length . . . . *Brachylagus*

1'. Anterior surface of first upper and lower premolars with two or more enamel re-entrant angles; auditory bulla smaller relative to size of cranium, width less than length of maxillary toothrow; palatal bridge more than 4.5 mm in length 2

2. Supraorbital process broad and more or less triangular in shape; interparietal bone indistinct, fused to parietals . . . . *Lepus*

2'. Supraorbital process relatively narrow and straplike, frequently in contact with, or fused to, parietal bone posteriorly; interparietal bone distinct, not fused to parietals . . . . *Sylvilagus*

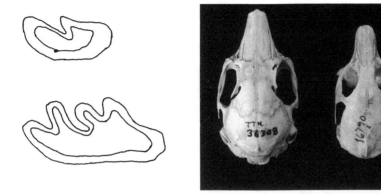

Fig. 14.—Left: drawings to scale of occlusal surface of first upper premolar on left side of *Brachylagus* (top) and *Sylvilagus* showing re-entrant angles in enamel. Right: dorsal views to scale of crania of *Brachylagus* (left, greatest length of skull, 51.8 mm) and *Ochotona*.

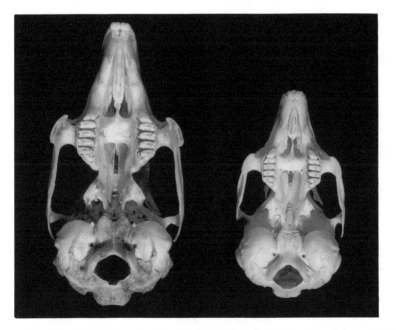

Fig. 15.—Ventral views to scale of skulls of *Sylvilagus* (left) and *Brachylagus* (greatest length of skull, 51.8 mm).

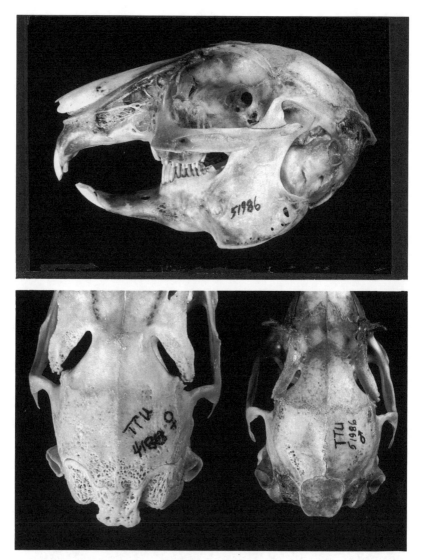

Fig. 16.—Top: lateral view of skull of *Sylvilagus* (greatest length of skull, 70.1 mm); note fenestrations in maxilla, small second upper incisor, and long diastema. Bottom: dorsal views to scale of posterior part of crania of *Lepus* (left) and *Sylvilagus*; note straplike supraorbital processes that are in contact with parietals and distinct interparietal in latter.

Order RODENTIA
Rodents

The order Rodentia comprises the largest and most diverse assortment of living mammals. It is represented by more than 30 families, about 490 genera, and some 1600 species. Rodents occupy environments from above the Arctic Circle to the tropics; they occur on all major land masses except Antarctica, and they have been introduced widely by humans. Most rodents are strictly terrestrial in habits, but the range in life style encompasses species that are fossorial and semiaquatic to those that are scansorial and even some true gliders. Eight families are native to North America north of México as follows (approximate number of species in parentheses): Aplodontidae (one); Sciuridae (65); Geomyidae (18); Heteromyidae (37); Castoridae (one); Muridae (80); Dipodidae (four); Erethizontidae (one). *Myocastor coypus,* representing the Myocastoridae, has been introduced into the region (from South America), as have the Old World murid genera *Mus* and *Rattus.*

The early ancestors of modern Rodentia are known from as far back in the fossil record as the late Paleocene. The order is characterized by having a single pair of ever-growing incisors in both upper and lower jaws, a dental formula that never exceeds 1/1, 0/0, 2/1, 3/3, total 22, and a distinct, broad diastema that separates the incisors from the cheekteeth. In contrast to lagomorphs, a baculum is present in males, enamel on the incisors is confined to the anterior and anterolateral surfaces, and maxillary fenestrations are lacking. Variation in number of premolars among rodents is shown in Table 3.

Table 3. Number of premolars in families of Rodentia (all North American rodents have a single pair of incisors above and below, lack canines, and have three molars above and below).

Family	Premolars	Family	Premolars
Aplodontidae	2/2	Muridae	0/0
Sciuridae	2-1/1	Dipodidae	1-0/0
Geomyidae	1/1	Erethizontidae	1/1
Heteromyidae	1/1	Myocastoridae	1/1
Castoridae	1/1		

Key to Families of Rodents

1. Skull flattened relative to length; upper cheekteeth (except peglike first premolar) with distinctive labial projection; auditory bulla flask-shaped, directed outward horizontally . . . . **Aplodontidae**

1'. Skull not noticeably flattened; no distinct labial projections on upper cheekteeth; auditory bulla not flask-shaped 2

2. Distinctive basioccipital pit; auditory canal of bulla long and directed both outward and upward; breadth of first upper cheektooth more than 9.5 mm at alveolus . . . . **Castoridae**

2'. No basioccipital pit; auditory canal of bulla short and not directed upward; breadth of first upper cheektooth less than 9.5 mm at alveolus 3

3. Length of maxillary toothrow more than 23 mm; infraorbital foramen large, more than a centimeter in diameter, larger than foramen magnum 4

3'. Length of maxillary toothrow less than 23 mm; infraorbital foramen smaller than foramen magnum (except in some Dipodidae), never more than a few millimeters in diameter 5

4. Paroccipital process long and well developed; last upper cheektooth (M3) largest in maxillary toothrow . . . . **Myocastoridae**

4'. Paroccipital process short and inconspicuous; first upper cheektooth (P4) largest in maxillary toothrow . . . . **Erethizontidae**

5. Mastoid part of bulla distinctly visible lateral to occipital region and usually from dorsal aspect of skull; infraorbital foramen on lateral surface of rostrum; cheekteeth always 4/4 6

5'. Mastoid bone inconspicuous; infraorbital foramen not on lateral surface of rostrum; cheekteeth 5/4, 4/4 (only in some sciurids), 4/3, or 3/3 7

6. Infraorbital foramen perforate; first upper cheek-tooth (P4) not constricted medially; last upper molar smallest tooth in row . . . . . **Heteromyidae**

6'. Infraorbital foramen not perforate; first upper cheek-tooth (P4) constricted medially, hourglass-shaped; last upper molar larger than other molars . . **Geomyidae**

7. Infraorbital foramen large, rounded, about size of external nares; separate, small foramen for nerve and blood vessels ventral to infraorbital foramen; vertical process of jugal extends upward almost to lacrimal . . . . **Dipodidae**

7'. Infraorbital foramen small hole to medium-sized slit, no separate foramen ventral to it; jugal without vertical process 8

8. Infraorbital foramen slitlike, situated above midline of rostrum; cheekteeth 3/3; no postorbital processes . . . . **Muridae**

8'. Infraorbital foramen a small hole, situated below midline of rostrum; cheekteeth 5/4 or 4/4; prominent postorbital processes . . . . **Sciuridae**

Aplodontidae (mountain beaver)

This monotypic family occurs only in the coastal regions and adjacent mountainous areas of the Pacific Northwest, from southern British Columbia to central California. The single modern species is *Aplodontia rufa*, the most primitive of living rodents. See Fig. 17.

Sciuridae (squirrels and allies)

Sciurids occur over much of the temperate and tropical parts of the world—in North and much of South America, Europe, most of Africa, and Asia southward through the Oriental Realm. Eight genera are found in the region here treated. Of these, three (*Ammospermophilus*, *Cynomys*, and *Tamiasciurus*) occur only in North America; the others are found also in the Palearctic Region.

Key to Genera of Sciuridae

1. Length of maxillary toothrow more than 18.5 mm;
   greatest length of skull more than 80 mm; temporal
   ridges distinct, elongated, V-shaped, uniting into
   sagittal crest posterior to terminus of zygomatic arch
   . . . . *Marmota*

1'. Length of maxillary toothrow less than 18.5 mm;
   greatest length of skull less than 80 mm; temporal
   ridge absent or indistinct (in all but a few of the larger
   species of *Spermophilus*) 2

2. Length of maxillary toothrow more than 14 mm; first
   upper cheektooth (P3) much broader than upper
   incisor . . . . *Cynomys*

2'. Length of maxillary toothrow less than 14 mm; P3
   present or absent, if present much narrower than
   upper incisor 3

3. Infraorbital canal absent; simple, rounded, infraorbital foramen piercing zygomatic plate . . . . *Tamias*

3'. Infraorbital canal, usually slitlike, passing between
   zygomatic plate and rostrum 4

4. Maxillary toothrows essentially parallel or even
   slightly divergent posteriorly; margin of interorbital
   region broadly V-shaped (best seen in ventral view)
   . . . . *Glaucomys*

4'. Maxillary toothrows convex, breadth across them
   greatest near midpoint; margin of interorbital region
   gently curved inward, sometimes notched medially,
   but not V-shaped 5

5. Zygomatic plate slanting upward from base to rostrum at angle of approximately 50 degrees; postorbital breadth about the same as, or only slightly greater than, interorbital breadth; premolars 1/1 or 2/1 (P3, if present, minute); enamel face of upper incisor roughened or rugose microscopically 6

5'. Zygomatic plate slating upward from base to rostrum at angle of approximately 40 degrees; postorbital breadth noticeably greater than interorbital breadth; premolars 2/1 (P3 well-developed peg in all except *Ammospermophilus*); enamel face of upper incisor smooth microscopically 7

6. Auditory bulla relatively large, with three dividing septa clearly visible externally; premolars 1/1; greatest length of skull less than 55 mm; anterior border of orbit opposite premolar or junction of premolar and first molar . . . . ***Tamiasciurus***

6'. Auditory bulla relatively small, bullar septa (if visible) numbering only two; premolars 1/1 or 2/1 (usually); greatest length of skull more than 55 mm; anterior border of orbit opposite second upper molar . . . . ***Sciurus***

7. Auditory bulla large, length about one and a half times that of maxillary toothrow; first premolar (P3) minute . . . . ***Ammospermophilus***

7'. Auditory bulla of normal size, length less than (or only slightly greater than) that of maxillary toothrow; first premolar (P3) a well-developed peglike tooth . . . . ***Spermophilus***

Note: Eastern and western chipmunks formerly were regarded as representing two distinct genera. Most authorities now regard these as subgenera, *Tamias* and *Neotamias*, respectively. The former has one upper premolar and the latter has two. Also, *Neotamias* has distinct, multiple longitudinal grooves, visible microscopically, on the enamel surface of the upper incisors, whereas *Tamias* has irregular, indistinct grooves.

Geomyidae (pocket gophers)

This fossorial family, together with the heteromyids, forms the superfamily Geomyoidea, a group limited to the New World. Three genera of geomyids are found in the region treated here; all occur also in México.

Key to Genera of Geomyidae

1. Anterior surface of upper incisor smooth; lower molars with transverse enamel plates on both anterior and posterior surfaces . . . . *Thomomys*

1'. Anterior surface of upper incisor grooved; lower molars with transverse enamel plate only on posterior surface 2

2. Anterior surface of upper incisor with single groove (unisulcate); transverse enamel plate only on anterior surface of first two upper molars . . *Cratogeomys*

2'. Anterior surface of upper incisors with two grooves (bisulcate); transverse enamel plate on both anterior and posterior surfaces of first two upper molars . . . . *Geomys*

Heteromyidae (pocket mice, kangaroo rats, and allies)

Pocket mice and their relatives occur in western North America, from southern Canada southward through Middle America into northern South America. There are six genera, all but one (*Heteromys*) occurring to the north of México, which are placed in three subfamilies: Heteromyinae (*Heteromys, Liomys*); Perognathinae (*Chaetodipus, Perognathus*); Dipodominae (*Dipodomys, Microdipodops*). None of these genera is monotypic. *Microdipodops* is placed in the Dipodominae following Ryan (1989).

Key to Genera of Heteromyidae

1. Anterior surface of upper incisor smooth; mastoid
   bulla barely visible in dorsal view, confined to area
   posterior to external auditory meatus . . . . *Liomys*

1'. Anterior surface of upper incisor grooved; mastoid
   bulla clearly visible in dorsal view and extending
   anteriorly to base of zygoma 2

2. Mastoid bulla markedly inflated, mastoid region
   much broader than zygomatic arches; interparietal
   much reduced, noticeably longer than broad . . . 3

2'. Mastoid bulla only moderately inflated, mastoid
   region about as broad as, or only slightly broader
   than, zygomatic arches; interparietal not conspicuously reduced, at least as broad as long 4

3. Greatest length of skull more than 32 mm; interparietal small but obvious, about half as broad as
   long; zygomatic process of maxilla broadly expanded
   . . . . *Dipodomys*

3'. Greatest length of skull less than 32 mm; interparietal
   a minute, longitudinal spicule; zygomatic process of
   maxilla only modestly expanded . . *Microdipodops*

4. Mastoid bulla more inflated, encroaching noticeably
   onto dorsal surface of skull and extending posteriorly
   beyond occiput; interparietal narrower than interorbital region . . . . *Perognathus*

4'. Mastoid bulla less inflated, mostly confined to lateral
   surface of skull and not extending posteriorly beyond
   occiput; interparietal as broad as, or broader than,
   interorbital region . . . . *Chaetodipus*

Castoridae (beavers)

In the Recent record, this family is represented only by
two closely related species, *Castor fiber* of the Palearctic
Region and *Castor canadensis* of the Nearctic. The latter
occurs from central Alaska and northern Canada southward

through much of the United States to northernmost México.
See Fig. 25.

Muridae (mice, rats, voles, and allies)

The distribution of the family Muridae coincides with that
of the order Rodentia. For many years, New World rats and
mice were regarded as belonging to the family Cricetidae,
which also had Old World representatives, but many authorities
(see especially Carleton and Musser, 1984) now regard them as
murids. The family Muridae, as reconstituted, contains 15
subfamilies, two of which occur naturally in North America:
Sigmodontinae (New World rats and mice) and Arvicolinae
(voles and their allies, which also are found in the Old World).
Additionally, species of two genera of the subfamily
Murinae (*Mus* and *Rattus*) have been introduced into North
America and are widely distributed in feral populations.
Furthermore, members of two other Old World subfamilies,
Cricetinae (hamsters) and Gerbillinae (gerbils) are common
household pets.

Key to Genera of Muridae

1. Molars cuspidate; cusps of first two upper molars
   arranged generally in three longitudinal rows
   (subfamily Murinae) 2

1'. Molars cuspidate, semiprismatic, or prismatic; if
    cuspidate, cusps of first two upper molars arranged
    in two longitudinal rows 3

2. Greatest length of skull more than 25 mm; upper
   incisor not notched on occlusal surface . . . ***Rattus***

2'. Greatest length of skull less than 25 mm; upper
    incisor distinctly notched posteriorly on occlusal
    surface (best seen in lateral view) . . . . ***Mus***

3. Zygomatic plate not extending anteriorly from zygomatic process of maxilla, leading edge straight or even concave; zygomatic notch poorly or not at all developed (best viewed from above); cheekteeth prismatic; maxillary toothrow terminating at about same plane as mandibular fossa or only slightly anteriorly (subfamily Arvicolinae) 4

3'. Zygomatic plate extending anteriorly parallel to rostrum from zygomatic process of maxilla; zygomatic notch moderately to well developed (best viewed from above); cheekteeth semiprismatic (*Neotoma*) to cuspidate; maxillary toothrow terminating well anterior of mandibular fossa (subfamily Sigmodontinae) 12

4. Greatest length of skull more than 40 mm; width of upper incisor more than 2.0 mm 5

4'. Greatest length of skull less than 40 mm; width of upper incisor less than 2.0 mm 6

5. Greatest length of skull more than 52 mm; width of upper incisor more than 3.0; anterior loop of lower first molar distinctly lobed and with deep re-entrant angles . . . . *Ondatra*

5'. Greatest length of skull less than 52 mm; width of upper incisor less than 2.2 mm; anterior loop of lower first molar rounded and with shallow re-entrant angles . . . . *Neofiber*

6. Upper incisor grooved anterolaterally; rostrum short, approximately as broad as long anterior to zygomatic arches . . . . *Synaptomys*

6'. Upper incisor smooth; rostrum longer than broad . 7

7. Cheekteeth rooted (brachyodont) in adults; nasals extending anteriorly beyond incisors 8

7'. Cheekteeth evergrowing (hypsodont); nasals not extending anteriorly beyond incisors 9

8. Palate terminating posteriorly in simple transverse shelf; cheekteeth small, upper toothrows somewhat divergent; lower molars with lingual re-entrant angles little if any deeper than those on labial side
. . . . *Clethrionomys*

8′. Palate terminating posteriorly with median spiny process; cheekteeth large, upper toothrows essentially parallel; lower molars with lingual re-entrant angles much deeper than those on labial side
. . . . *Phenacomys*

9. Squamosal bone with prominent "knob" or "peg" extending anterolaterally into orbit; first lower molar with seven closed triangles between anterior and posterior loops . . . . *Dicrostonyx*

9′. Squamosal bone without prominent extensions; lower first molar with six or fewer closed triangles between anterior and posterior loops 10

10. Skull massive, with broadly divergent zygoma having a broad flange near jugo-maxillary suture; upper toothrows strongly divergent posteriorly; first lower molar with three closed triangles . . . . . *Lemmus*

10′. Skull not massive, with more-or-less evenly rounded zygoma that are only modestly or not at all broadened at jugo-maxillary junction; upper toothrows only slightly divergent posteriorly; first lower molar with more than three closed triangles 11

11. Skull small, greatest length usually less than 22 mm; internal surface of anterior loop of first lower molar smooth . . . . *Lemmiscus*

11′. Skull larger, greatest length usually 25 mm or more; internal surface of anterior loop of first lower molar distinctly notched with small re-entrant fold
. . . . *Microtus*

12. Cheekteeth semiprismatic, with deep re-entrant angles; greatest length of skull more than 40 mm
. . . . *Neotoma*

12'. Cheekteeth cuspidate; greatest length of skull less
than 40 mm .13

13. Temporal ridge strongly developed; zygomatic
notch deep .14

13'. Temporal ridge absent; zygomatic notch shallow. .15

14. Cheekteeth robust, breadth of first upper molar about
same as breadth across incisive foramina; maxillary
toothrows divergent; paroccipital process relatively
well developed, projecting below level of occipital
condyle . . . . *Sigmodon*

14'. Cheekteeth small, breadth of first upper molar less
than breadth across incisive foramina; maxillary
toothrows convergent posteriorly; paroccipital
process small, not projecting below level of occipital
condyle . . . . *Oryzomys*

15. Upper incisor with distinct median groove on
anterolateral surface . . . . *Reithrodontomys*

15'. Upper incisor smooth 16

16. Greatest length of skull less than 20 mm; length of
maxillary toothrow less than 3.2 mm . . . *Baiomys*

16'. Greatest length of skull more than 20 mm; length of
maxillary toothrow more than 3.2 mm 17

17. Cusps on first two molars both above and below well
developed, especially labially, projecting outward in
unworn teeth; coronoid process high and elongated
. . . . *Onychomys*

17'. Cusps on molars normally developed, not projecting
outward; coronoid process small, not extending above
articular process 18

18. Posterior palatine foramen opposite level of second
upper molar; partly worn occlusal surface of first
upper molar complex, with five subtriangular islets of
dentine isolated by enamel ridges . . . *Ochrotomys*

18′. Posterior palatine foramen opposite junction of first and second upper molars; partly worn occlusal surface of first upper molar (and other molars) less complex, with no or few islets of dentine isolated by enamel ridges **19**

19. Small supraorbital shelf present; molars lacking accessory ridges and cusplets; larger than any sympatric species of *Peromyscus* in southeastern United States . . . . *Podomys*

19′. Supraorbital shelf absent; at least first molar normally with some accessory ridges and cusplets; sympatric species smaller than *Podomys* . . . . . *Peromyscus*

Notes: The generic as opposed to subgeneric status of several groups of American *Microtus*, namely *Arvicola, Pitymys*, and *Pedomys*, has been widely debated in the literature. We follow Carleton and Musser (1984) in regarding all as *Microtus*, as we do in retaining *Arborimus* as a subgenus of *Phenacomys*. Carleton and Musser also are our authorities for use of *Lemmiscus* in place of *Lagurus* for the sagebrush vole (Klingener, 1984).

Dipodidae (jumping mice, jerboas, and allies)

Jumping mice formerly were regarded as belonging to a separate family, Zapodidae, with four genera, two in North America and two in the Palearctic. More recently, this group has been relegated to subfamilial status under Dipodidae (Klingener, 1984).

Key to Genera of Dipodidae

1. Cheekteeth 4/3; first upper molar with four labial reentrant folds . . . . *Zapus*

1′. Cheekteeth 3/3; first upper molar with three labial reentrant folds . . . . *Napaeozapus*

Erethizontidae (American porcupines)

"Porcupines" are found in two different (and only distantly related) families, one in the Old World and one in the New. The family of the Americas, Erethizontidae, contains four genera, only one of which, the monotypic *Erethizon*, occurs to the north of the tropics. The one species, *E. dorsatum,* is found from Alaska and Canada southward through much of the United States to northern México. See Fig. 25.

Myocastoridae (nutria or coypu)

This monotypic family is native to southern South America. The one species, the semiaquatic *Myocastor coypus,* has been introduced into the southern and western United States, and substantial feral populations now occur in some places, particularly the coastal region of the Gulf of Mexico. See Fig. 35.

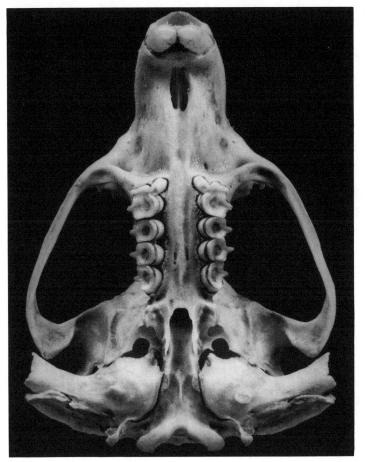

Fig. 17.—Ventral view of cranium of *Aplodontia* (greatest length of skull, 70.5 mm).

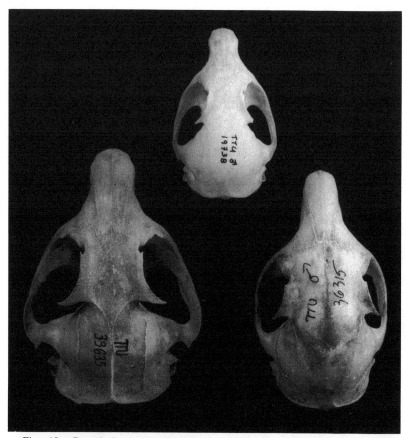

Fig. 18.—Dorsal views to scale of crania of *Cynomys* (lower left), *Spermophilus* (above), and *Sciurus*. Greatest length of skull of the *Cynomys* is 65.1 mm.

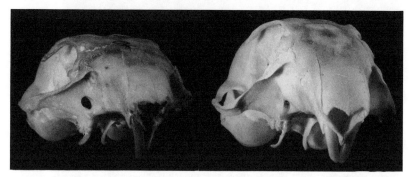

Fig. 19.—Anterolateral views to scale of crania of *Tamias* (left) and *Spermophilus*. Note the infraorbital foramen, which pierces the zygomatic plate in *Tamias* as opposed to the canal between the plate and the rostrum in *Spermophilus*.

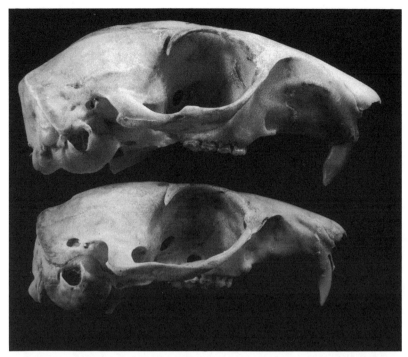

Fig. 20.—Lateral views of crania of *Sciurus* (above) and *Spermophilus*. Note angle of zygomatic plate.

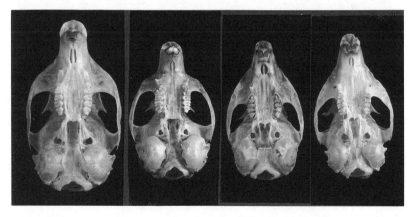

Fig. 21.—Ventral views to scale of crania of *Tamiasciurus*, *Glaucomys*, *Ammospermophilus*, and *Tamias*, left to right. Greatest length of skull of the *Tamiasciurus* is 49.1 mm.

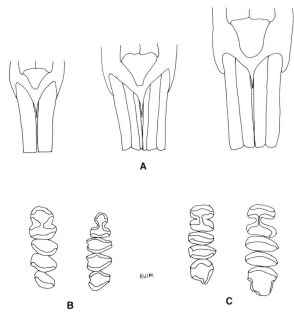

Fig. 22.—A, drawings to scale of anterior surfaces of upper incisors of *Thomomys, Geomys,* and *Cratogeomys,* left to right; B, drawings to scale of occlusal surfaces of left lower cheekteeth of *Geomys* (left) and *Thomomys*; C, drawings to scale of occlusal surfaces of left upper cheekteeth of *Geomys* (left) and *Cratogeomys*.

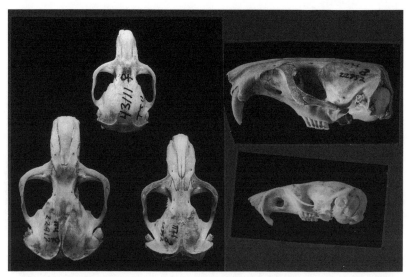

Fig. 23.—Left: dorsal views to scale of crania of pocket gophers (*Cratogeomys,* left, *Thomomys,* above, and *Geomys*); greatest length of skull of the *Cratogeomys* is 48.8 mm. Right: lateral views to scale of crania of *Geomys* (above) and *Liomys*.

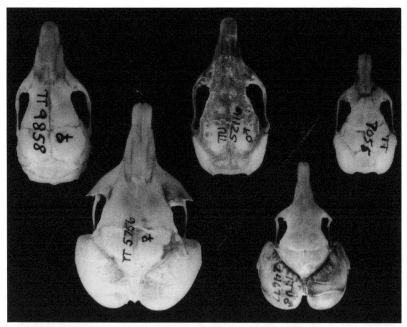

Fig. 24.—Dorsal views to scale of crania of heteromyids. Top: *Liomys, Chaetodipus,* and *Perognathus,* left to right. Bottom: *Dipodomys* (left) and *Microdipodops.* Greatest length of skull of the *Liomys* is 30.7 mm.

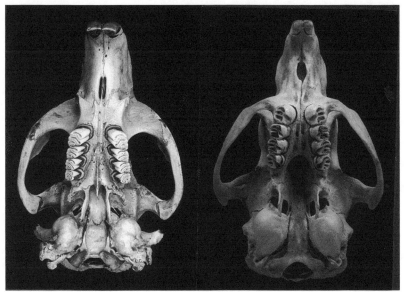

Fig. 25.—Ventral views of crania of beaver (*Castor*), left, and porcupine (*Erethizon*); photographs not to same scale.

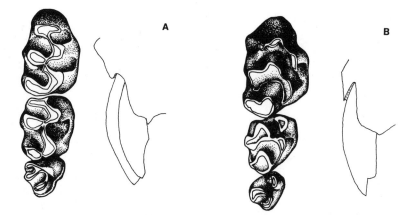

Fig. 26.—A, diagram of left upper molars and lateral outline of upper incisor of *Reithrodontomys*; B, diagram of left upper molars and lateral outline of upper incisor in *Mus* (note notch in incisor and three rows of cusps on first molar). Modified after Jones *et al.* (1983).

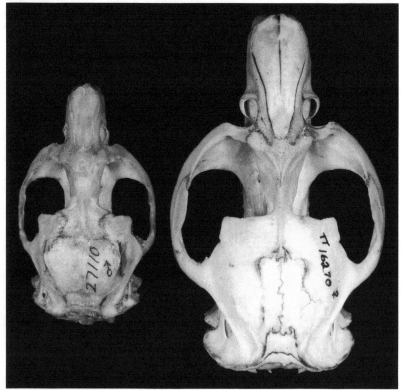

Fig. 27.—Dorsal views to scale of crania of *Neofiber* (left) and *Ondatra* (greatest length of skull, 64.5 mm).

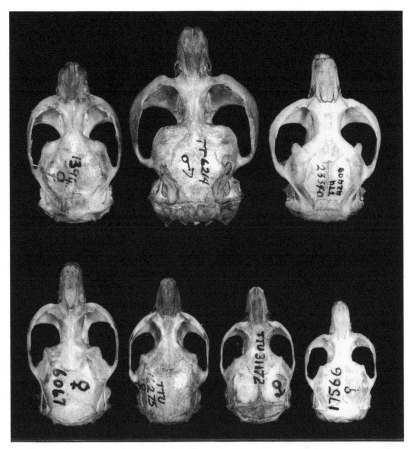

Fig. 28.—Dorsal views to scale of crania of seven genera of microtines. Top: *Synaptomys, Lemmus,* and *Dicrostonyx*, left to right. Bottom: *Microtus, Phenacomys, Clethrionomys,* and *Lemmiscus*, left to right. Greatest length of skull of the *Lemmus* is 35.3 mm.

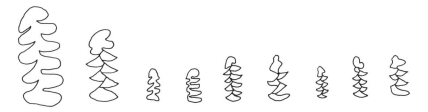

Fig. 29.—Drawings (to scale) of occlusal surface of first lower left molar of nine genera of microtines (anterior up, lingual right). Left to right: *Ondatra, Neofiber, Clethrionomys, Phenacomys, Dicrostonyx, Lemmus, Lemmiscus, Microtus,* and *Synaptomys.*

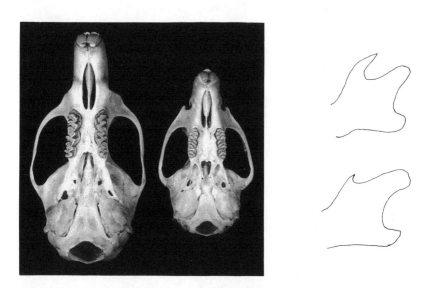

Fig. 30.—Ventral views of crania of *Neotoma* (left) and *Sigmodon* (to scale), and drawings (to scale) of posterior part of mandibles of *Onychomys* (top) and *Peromyscus* showing the long coronoid process of the former. Greatest length of skull of the *Neotoma* is 48.0 mm.

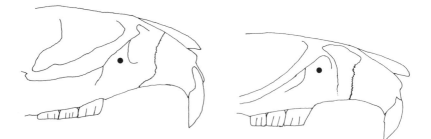

Fig. 31.—Drawings to scale of lateral views of crania of *Neofiber* (left) and *Neotoma* indicating differences in zygomatic plate (marked with dot) and thus development of zygomatic notch.

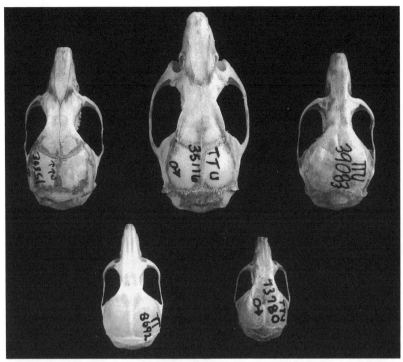

Fig. 32.—Dorsal views of crania of five genera of murids. Top row: *Onychomys, Oryzomys* (greatest length of skull, 32.9 mm), and *Ochrotomys* left to right. Bottom: *Reithrodontomys* (left) and *Baiomys*.

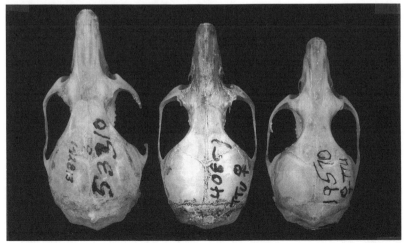

Fig. 33.—Dorsal views of crania of *Podomys* (left) and two sympatric species of *Peromyscus* (*gossypinus*, center, and *leucopus*). Greatest length of skull of the *Podomys* is 31.2 mm.

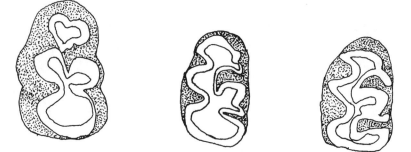

Fig. 34.—Drawings to scale of left first upper molar of *Podomys, Peromyscus* (*leucopus*), and *Ochrotomys*, left to right.

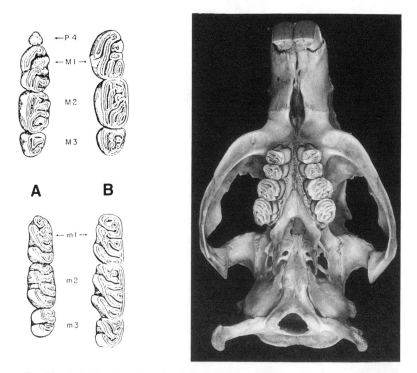

Fig. 35.—Left: drawings of occlusal views to scale of upper and lower right cheekteeth of *Zapus* (A) and *Napaeozapus* (B), after Krutzsch (1954). Right: ventral view of cranium of *Myocastor* (greatest length of skull, 119.8 mm), at slightly oblique angle to emphasize right paroccipital process.

Order CARNIVORA
Carnivores

Carnivores first appear in the fossil record in Paleocene times, some 60 million years ago. Ten modern families, including the aquatic "pinnipeds," usually are recognized, of which five (Canidae, Ursidae, Procyonidae, Mustelidae, and Felidae) are included here. Four of these families have broad distributions over much of the world; the Procyonidae, however, is restricted to the Americas, except for the lesser panda of southeastern Asia. As a group, terrestrial carnivores are found on all continents except Antarctica and on Australia (where the wild dog, or dingo, probably was introduced by early man); aquatic carnivores, however, occur along the coasts of those two continents and along coastlines of many oceanic islands. Recent representatives of the order are classified in about 100 genera and total some 270 species.

Terrestrial carnivores occupy a variety of ecological settings from the north polar region to tropical rainforests. As the ordinal name implies, most typically eat flesh of other vertebrates, but some species regularly take fruits, nuts, and other vegetable matter, as well as invertebrates. All North Ameircan representatives have well-developed canines and most have a specialized pair of shearing teeth, the carnassials, formed by the fourth upper premolar and the first lower molar. Carnassials are best developed in felids, but are well developed in canids and mustelids. They are reduced secondarily in ursids and procyonids. All North American carnivores have a well-developed baculum except the felids, in which it is small.

Key to Families of Carnivores

1. Molars 1/1, upper molar much reduced; premolars 2/2 or 3/2 . . . . **Felidae**

1′. Molars 1/2 or more, upper molar (or first upper molar if more than one present) a large and robust tooth; premolars variable but never 2/2 or 3/2 2

2. Molars 1/2 to 2/2; alisphenoid canal absent 3

2′. Molars 2/3; alisphenoid canal present 4

3. Molars 1/2; total of 32 to 38 teeth; carnassial pair well developed . . . . **Mustelidae**

3′. Molars 2/2; total of 40 teeth; carnassial pair secondarily reduced except in *Bassariscus* . . **Procyonidae**

4. Maxillary toothrow more or less straight; cheekteeth flattened; last upper molar largest postcanine tooth above . . . . **Ursidae**

4′. Maxillary toothrow noticeably divergent at level of last premolar; cheekteeth with sharpened cusps; last upper premolar largest postcanine tooth above . . . . **Canidae**

Canidae (wolves, coyote, and foxes)

Canids occur from arctic habitats to deserts and tropical areas. The distribution is broadest among terrestrial families of the order Carnivora, slightly more extensive than that of Mustelidae. Worldwide, there are 12 modern genera that contain a total of about 34 species. Four genera occur in North America, *Canis* (three species), *Vulpes* and *Urocyon* (each with two species), and the monotypic arctic fox, *Alopex*. Of these, only *Urocyon* is limited to the New World. All North American canids have the same dental formula, 3/3, 1/1, 4/4, 2/3, total 42. See Fig. 36.

Key to Genera of Canidae

1. Temporal ridge distinctly lyre-shaped; depression in frontal bone just medial to postorbital process; truncated indentation on lower margin of dentary anterior to angular process . . . . *Urocyon*

1′. Temporal ridge not lyre-shaped; no frontal depression; lower margin of dentary smooth 2

2. Anterior palatine foramen short, not extending
   posteriorly beyond midpoint of canine; rostrum
   relativley short and broad, rising rather abruptly to
   braincase in lateral view . . . . *Alopex*

2'. Anterior palatine foramen long, extending beyond
    midpoint of canine and usually beyond that tooth;
    rostrum relatively long and narrow, rising gradually
    to braincase in lateral view 3

3. Greatest length of skull less than 200 mm; postorbital process well developed, postorbital breadth
   noticeably less than interorbital breadth . . . *Vulpes*

3'. Greatest length of skull more than 200 mm; post-
    orbital process weakly developed, postorbital
    breadth about same as, or greater than, interorbital
    breadth . . . . *Canis*

Note: Some breeds of domestic dogs (*Canis familiaris*) cannot be
identified through characters used in the foregoing key.

Ursidae (bears)

Modern bears are regarded as represented by six genera,
the giant panda (*Ailuropoda*) in one subfamily and the
remainder in the subfamily Ursinae, members of which
occur (or occurred within historic time) throughout North
America and Eurasia, including southeastward onto the
Malayan Peninsula, in mountainous South America, and in
the Atlas Mountains of North Africa. Three species of the
genus *Ursus* are found in North America—polar bear, brown
or grizzly bear, and black bear. These once were regarded as
representing three distinct genera, *Thalarctos, Ursus,* and
Euarctos, respectively, but now are regarded at best as subgenera. Ursids have four premolars both above and below,
but the first three frequently are rudimentary and sometimes
lost. See Fig. 37.

Procyonidae (raccoons and allies)

Aside from the lesser panda, *Ailurus*, of southeastern
Asia, Recent procyonids are restricted to the New World,

where six genera encompassing 18 species are known; most are tropical or subtropical in distribution. Three genera are treated here, each containing a single species in the area of coverage. All procyonids treated have the same dental formula, 3/3, 1/1, 4/4, 2/2, total 40.

Key to Genera of Procyonidae

1. Bony palate terminating just posterior to last upper molar; upper molars transversely elongated, much broader than long . . . . *Bassariscus*

1′. Bony palate terminating far posterior to last upper molar (posterior to optic foramen); upper molars not transversely elongated, about as broad as long . . 2

2. Rostrum noticeably elongated and greatly compressed laterally, width anteriorly much less than postorbital width; conspicuous gap between canine and first upper premolar . . . . *Nasua*

2′ Rostrum only moderately elongated, not conspicuously compressed laterally, about same width as postorbital region; no conspicuous gap between canine and first upper premolar . . . . . . *Procyon*

Mustelidae (weasels, skunks, and allies)

There is great variation among mustelids, from the wolverine, which approaches a small bear in size, down to the least weasel, some individuals of which are no larger than a robust mouse. The family is found worldwide except on Antarctica, Australia, Madagascar, and most oceanic islands, but is most common, diverse, and widely distributed in the Holarctic Region. There are 23 Recent genera comprised of some 63 species. Nine genera and 17 species are found in North America north of México. Of these, the three genera of skunks (*Conepatus, Mephitis,* and *Spilogale*) and the badger (*Taxidea*) are restricted to the New World, whereas *Enhydra, Gulo, Lutra, Martes,* and *Mustela* are known from at least parts of the Old World as well. See Table 4 for dental formulae of genera. In mustelids, adult males normally average

Table 4. Dental formulae of nine genera of Mustelidae.

Genus	Incisors	Canines	Premolars	Molars	Total
Enhydra	3/2	1/1	3/3	1/2	32
Conepatus	3/3	1/1	2/3	1/2	32
Mephitis	3/3	1/1	3/3	1/2	34
Mustela	3/3	1/1	3/3	1/2	34
Spilogale	3/3	1/1	3/3	1/2	34
Taxidea	3/3	1/1	3/3	1/2	34
Lutra	3/3	1/1	4/3	1/2	36
Gulo	3/3	1/1	4/4	1/2	38
Martes	3/3	1/1	4/4	1/2	38

10 to 15 percent larger than females, and tend to have better developed cranial crests.

Key to Genera of Mustelidae

1. Incisors 3/2; posterior lacerate foramen large (about 8 mm in diameter) and rounded . . ***Enhydra***

1'. Incisors 3/3; posterior lacerate foramen small, usually inconspicuous 2

2. Premolars 4/3; infraorbital foramen large, roughly oblong, greatest diameter more than 8 mm; post-orbital process well developed . . . . ***Lutra***

2'. Premolars 2/3, 3/3, or 4/4; infraorbital foramen ovoid, triangular, or oblong, diameter much less than 8 mm; postorbital process lacking to moderately well developed (*Martes, Mustela*) 3

3. Greatest length of skull more than 130 mm; teeth large and robust, first lower molar more than 18 mm in length . . . . ***Gulo***

3'. Greatest length of skull less than 130 mm; teeth large to moderate in size, first lower molar less than 15 mm in length 4

4. Skull roughly triangular in outline, broadest posteriorly; upper molar large and triangular in shape; auditory bulla large, noticeably inflated ventrally . . *Taxidea*

4'. Skull not triangular in outline; upper molar quadrate or transversely elongate in shape; auditory bulla not noticeably inflated ventrally 5

5. Premolars 4/4 . . . . *Martes*

5'. Premolars 2/3 or 3/3 6

6. Upper molar transversely placed and dumbell-shaped; auditory bulla prominent and elongated; bony palate extending well beyond terminus of toothrow . . . . *Mustela*

6'. Upper molar quadrate or subquadrate; auditory bulla flattened and not elongated; bony palate extending only slightly beyond terminus of toothrow 7

7. Greatest length of skull less than 65 mm; small postorbital process; first upper premolar (P2) well developed, extending well below cingulum of second premolar . . . . *Spilogale*

7'. Greatest length of skull more than 65 mm; no postorbital process; first upper premolar minute or lacking 8

8. Premolars 2/3; upper molar about as long as broad; mastoid region not concave ventrally . . *Conepatus*

8'. Premolars 3/3; upper molar broader than long; mastoid region noticeably concave ventrally . . . . *Mephitis*

Felidae (cats)

Cats are found throughout the world in boreal, temperate, and tropical environments, but are absent in polar regions. Thirty-seven species are recognized worldwide; six have been recorded in North America to the north of México. There is disagreement in the literature concerning the number of felid genera. We here treat the North American species all as members of the genus *Felis,* but three different genera have been recognized by some mammalogists. The classical distinction between the two best known of these, *Felis* and *Lynx,* is that premolars are 3/2 in the former and 2/2 in the latter. See Fig. 42.

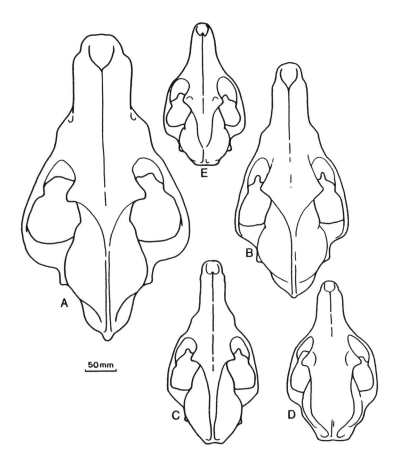

Fig. 36.—Drawings to scale of three genera of canids: A, *Canis (lupus)*; B, *Canis (latrans)*; C, *Vulpes (vulpes)*; D, *Urocyon*; E, *Vulpes (velox)*. After Jones *et al.* (1983).

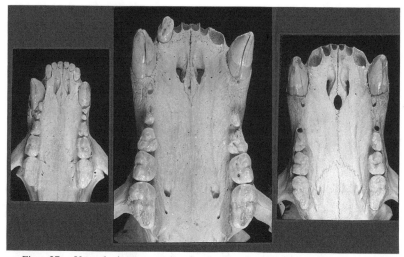

Fig. 37.—Ventral views to scale of palatal region of three subgenera of *Ursus* (*Euarctos, Ursus,* and *Thalarctos*, left to right). Note variable number of premolars and large, flattened last molar.

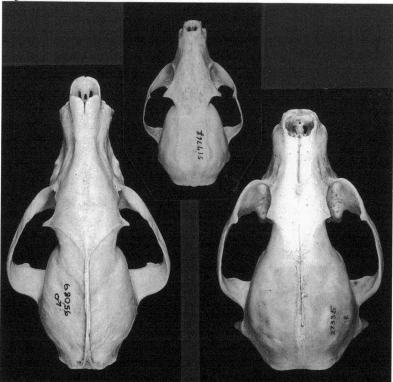

Fig. 38.—Dorsal views to scale of crania of procyonids (*Nasua, Bassariscus,* and *Procyon*, left to right). Greatest length of skull of the *Bassariscus* is 77.2 mm).

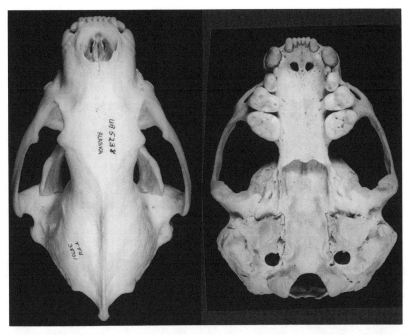

Fig. 39.—Dorsal view of the cranium of *Gulo* and ventral view of the cranium of *Enhydra*, to same scale. Greatest length of skull of the latter is 140.9 mm.

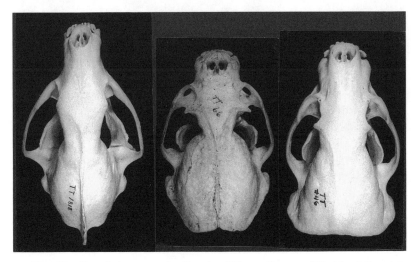

Fig. 40.—Dorsal views to scale of crania of *Martes, Lutra,* and *Taxidea*, left to right. Note distinct triangular outline of latter. Greatest length of skull of the *Taxidea* is 115.7 mm.

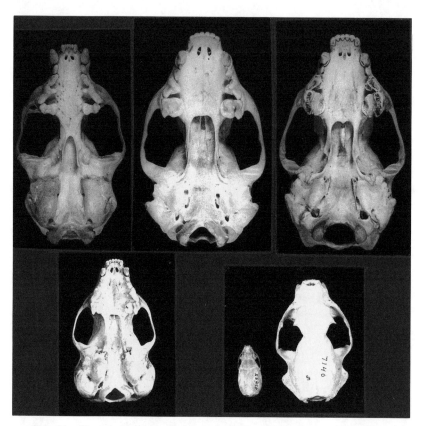

Fig. 41.—Ventral views to scale of crania of *Mustela* (greatest length of skull, 67.5 mm), *Mephitis, Conepatus* (left to right, above), and *Spilogale* (left, below). Dorsal views of crania (right, below) of smallest and largest North American *Mustela* (*nivalis* and *nigripes*).

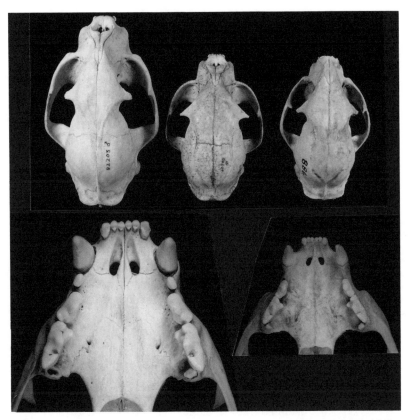

Fig. 42.—Top: dorsal views to scale of crania of three species of *Felis* (*pardalis, wiedii,* and *yagouaroundi,* from left to right); note shortened rostrum. Bottom: ventral views to scale of upper toothrows of two species of *Felis* (*concolor,* left, and *rufus*); note differing number of premolars and small molar.

Order ARTIODACTYLA
Artiodactyls

Artiodactyls, or even-toed ungulates, first appear in the fossil record in the early Eocene. The paleontological history of the group is reasonably well known. Recent species, of which there are about 185, arranged in nine families and representing almost 80 genera, occur worldwide except on Antarctica, Australia, and a number of oceanic islands including New Zealand. However, translocation and purposeful or accidental release has resulted in wide dispersal of many species in the past several centuries. Exotic and domesticated ungulates are not covered in the present work.

Four families of native artiodactyls are found in North America to the north of México—Antilocapridae (one genus), Bovidae (four), Cervidae (four), and Dicotylidae (one). Among the more trenchant characters of artiodactyls are the following: baculum absent, main axis of the foot passes between third and fourth digits (paraxonic), which are subequal and larger than such lateral digits as may be present; preorbital part of skull elongated; teeth brachyodont or hypsodont, with crown pattern bunodont or (usually) selenodont; lower canines incisiform (except Dicotylidae); premolars not fully molariform; ruminant digestive system except in dicotylids; terminal phalanges encased in hoofs. Because there are only 10 genera of North American artiodactyls, we treat them in a single key, with internal reference within the key as to breaks at the familial level.

Key to Genera of Artiodactyla

1. No appendages on frontal bones; upper incisors present; upper canine well developed (Dicotylidae) . . . . *Tayassu*

1'. Appendages (horns or antlers) present on frontal bones at least in males; upper incisors absent; upper canine absent or poorly developed 2

2. Frontal appendages nondeciduous, present in both sexes; lacrimal articulating with nasal, rostral fenestration absent (Bovidae) 3

2'. Frontal appendages deciduous, absent in females of some species; lacrimal not articulating with nasal, rostral fenestration present 6

3. Skull massive, greatest length more than 350 mm; length of maxillary toothrow more than 110 mm

   4

3'. Skull moderate in size, greatest length less than 350 mm; length of maxillary toothrow less than 110 mm

   5

4. Horn broad and flattened at base, decurved proximately, arising well internal to orbit; bone surrounding orbit distinctly tubular in shape . . . . . *Ovibos*

4'. Horn circular at base, not flattened, not decurved, arising at about same level as orbit; bone surrounding orbit not strongly tubular in shape . . . . . . *Bison*

5. Lacrimal pit absent; horn only slightly curved, less than 150 mm in circumference at base . . *Oreamnos*

5'. Lacrimal pit present; horn strongly curved, more than 150 mm in circumference at base . . . . *Ovis*

6. Frontal appendages composed of fused hairs surrounding bony core, arising from skull above posterior plane of orbits; rostral fenestration long and narrow (Antilocapridae) . . . . . . *Antilocapra*

6'. Frontal appendages composed of bone, arising from skull well posterior of orbits; rostral fenestration nearly as broad as long (Cervidae) 7

7. Antlers strongly palmated; nasal short, length much less than from its tip to anterior terminus of premaxilla . . . . *Alces*

7'. Antlers partially palmated (*Rangifer*) or not palmated; nasal relatively long, its length much more than from tip to anterior terminus of premaxilla

   8

8. Antlers usually present in both sexes, with palmated brow tine extending forward over rostrum from one or both main beams; premaxilla not in contact with nasal . . . . ***Rangifer***

8'. Antlers present only in males, lacking palmated brow tine; premaxilla in contact with nasal 9

9. Upper canine present; internal nares not divided by vomer . . . . ***Cervus***

9'. Upper canine absent; internal nares divided vertically by vomer . . . . ***Odocoileus***

Dicotylidae (peccaries)

Peccaries are restricted to the New World, occuring from the southwestern United States to southern South America. There are three Recent species, each in a monotypic genus. *Tayassu tajacu* is the one representative in temperate North America. Feral hogs (family Suidae), the skulls of which superficially resemble those of peccaries, can be distinguished easily from them because, among other things, they have incisors 3/3 instead of 2/3 and premolars 4/4 instead of 3/3. Feral hogs are found mostly in the Southeast.

We follow recent workers in use of the generic name *Tayassu* for the collared peccary, and the familial name Dicotylidae (see Jones *et al.*, 1992). The cheekteeth of peccaries are bunodont in comparison to the selenodont cheekteeth of other native artiodactyls.

Cervidae (deer and allies)

Members of this family are distributed over most of the geographic range of the order, but a number of species have been widely introduced at places where they did not occur natively. There are 17 modern genera that encompass about 36 species. Of the four genera that occur in temperate North America, three also are found in the Old World. *Odocoileus,* with two species, is restricted to the Western Hemisphere.

Antilocapridae (pronghorn)

Although it has a rich fossil record extending back to early Miocene times, this now is a monotypic family, restricted to the plains of western North America. The single species is *Antilocapra americana*. Some recent authorities have regarded antilocaprids as a subfamily within Bovidae, whereas others have considered them as cervids.

Bovidae (bison, cattle, and allies)

Bovids occur natively throughout most of Africa, over much of Eurasia, and in temperate and northern North America. Of the four genera covered in this text, the mountain goat, *Oreamnos* (monotypic), occurs only in the mountains of western North America. The others, *Bison, Ovibos* (monotypic), and *Ovis*, all occur also in Eurasia. A single species of *Bison* and two of *Ovis* are known from the New World. In many recent references, the bison has been referred to the genus *Bos* (see Jones *et al.*, 1992).

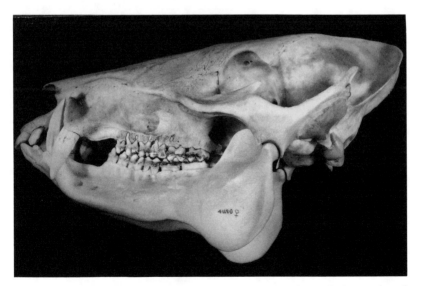

Fig. 43.—Lateral view of skull of *Tayassu*. Note presence of upper incisors, enlarged canines, modest diastema, and incomplete postorbital bar. Greatest length of skull is 256.0 mm.

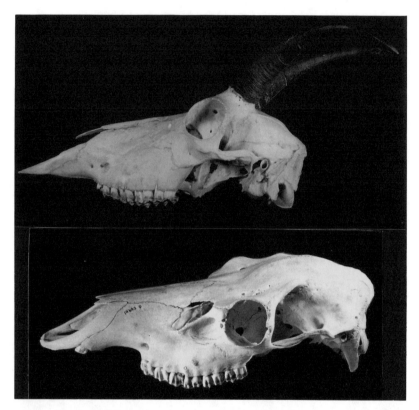

Fig. 44.—Lateral views of crania, not to same sale, of *Oreamnos* (top) and *Cervus*. By way of example, note that the premaxilla is in contact with the nasal in *Cervus* but not in *Oreamnos*.

GLOSSARY

accessory cusp—A small cusp usually situated peripheral to main biting or crushing surface of a tooth.

alveolus—Socket in jawbone that receives root(s) of tooth or contains base of ever-growing tooth.

alisphenoid canal—Passageway in the base of alisphenoid bone through which a blood vessel passes. The anterior and posterior openings of the canal are the third and fourth foramina at the posterior base of the orbit in Fig. 45.

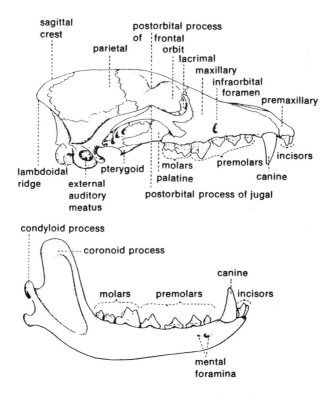

Fig. 45.—Drawings (with many features labeled) of lateral views of cranium and mandible of *Canis* (after Bekoff, 1977).

angular process—Posteroventral bony projection of dentary ventral to condyloid (articular) process. See Fig. 46.

anterior palatine foramen—See incisive foramen.

antler—Branched (usually), bony head ornament of frontal bone found in cervids, often only in males, covered with skin (velvet) during growth; shed annually.

articular process (or condyloid process)—Posterior bony projection of dentary that supports the articular condyle (point of articulation between lower jaw and cranium). See Figs. 45 and 46.

articulate—To join or connect two adjacent bones.

auditory bulla—Bony capsule enclosing middle ear; when formed by tympanic bone termed tympanic bulla. See Fig. 46.

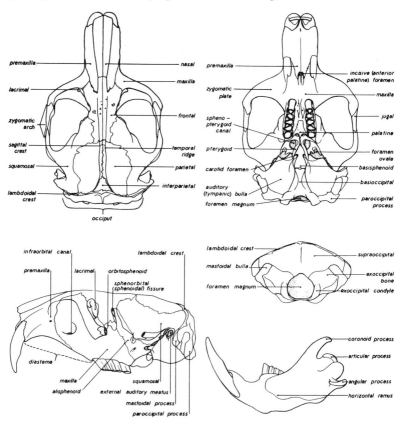

Fig. 46.—Drawings (with many features labeled) of dorsal and ventral views (above), lateral view (left, below), and posterior view (top right, below) of cranium, and lateral view of mandible (bottom right, below) of *Thomomys* (modified after Hall, 1955).

auditory canal—Bony tubular passage between tympanic membrane and external auditory meatus.

baculum—Sesamoid bone (*os penis*) in penis of males of certain mammalian groups.

basioccipital—Unpaired bone at base of occipital complex. See Fig. 46.

brachyodont—Pertaining to a low-crowned tooth; such teeth are rooted.

braincase—Posterior portion of cranium; part that encloses and protects the brain.

browtine—First tine above the base on an antler.

bunodont—Low-crowned, rectangular, grinding tooth, typical of omnivores.

canine—One of four basic kinds of mammalian teeth; anteriormost in maxilla (and counterpart of dentary); frequently elongate, unicuspid, and single-rooted; never more than one per quadrant. See Figs. 45 and 47.

caniniform—Having the general shape of a canine.

carnassial pair—Pair of large bladelike teeth (last upper premolar and first lower molar) that occlude with scissorlike (shearing) action, possessed by most modern carnivores.

cheekteeth—Collectively, postcanine teeth (premolars and molars).

cingulum—Enamel shelf bordering margin(s) of a tooth (cingulid used for shelf of lower teeth). See Fig. 49.

condyloid process—See articular process.

coronoid process—Posterior bony projection of dentary anterodorsal to articular process. See Figs. 45 and 46.

cranium—Collectively, bones that form upper part of skull (contains upper teeth and braincase); lower part of skull is the mandible.

Cretaceous—See geologic time.

cusp—Point, projection, or bump on crown (chewing surface) of a tooth.

cuspidate—Presence of cusps on a tooth.

cusplet (or secondary cusp)—A small cusp.

deciduous dentition (or milk teeth)—Juvenile teeth, those that appear first in lifetime of a mammal, consisting (if complete) of incisors, canines, and premolars; generally replaced by adult (permanent) dentition.

decurve—To curve downward.

dental formula—A numerical representation (shorthand method used by mammalogists) of the kind (incisor, canine, premolar, and molar) and number of each kind of tooth on one side of the upper and one side of the lower mammalian jaws.

dentary—Bone of lower jaw, forming half of mandible.

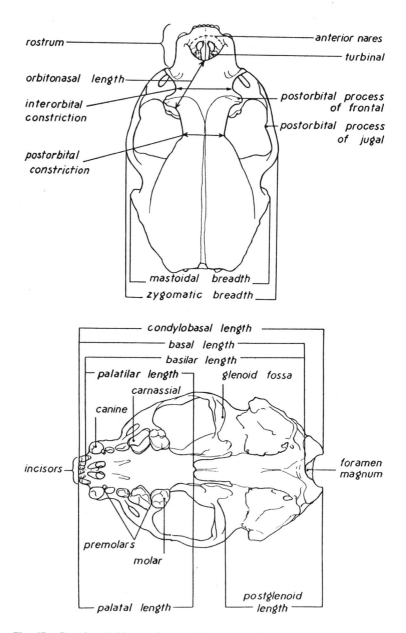

Fig. 47.—Drawings (with some features labeled and various measurements shown) of dorsal and ventral views of cranium of *Lutra* (modified after Hall, 1955).

dentine—Calcareous material, harder than bone but softer than enamel, which makes up much or most of a tooth.

diastema—A gap or space in jaw between teeth. Used most often to denote gap between incisors and cheekteeth in lagomorphs and rodents. See Fig. 46.

enamel—layer of material (usually outermost layer) covering a tooth; hardest substance in the body.

enamel plate—A segment or portion of a tooth that is heavily invested with enamel (for example, on pocket gopher teeth).

Eocene—See geologic time.

external auditory meatus—Round bony oriface or opening that is covered by eardrum. See Figs. 45 and 46.

fenestration—Opening; in current work applied to specialized openings in crania of lagomorphs and cervids.

flange—A laterally compressed or flattened portion of bone that increases the surface area.

foramen—Opening in bone through which passes nerves, blood vessels, or muscles.

foramen magnum—Large opening on the posterior of a cranium through which passes the spinal cord. See Figs. 46 and 47.

fossa—A shallow depression on surface of bone.

frontal—Paired bone of cranium, near orbit, situated posterior to nasal and anterior to parietal. See Fig. 46.

frontal appendages—Bone growth (horns or antlers) arising from frontal bones.

geologic time—Mammals arose in the Mesozoic Era, which began some 230 million years ago. The following Cenozoic Era, the "Age of Mammals" was the time of evolution and radiation of major modern mammalian groups. The Cenozoic is divided into two periods, Tertiary (beginning about 63 million years ago and continuing until about two million years ago) and Quaternary (approximately two million years ago to present). Subdivisions of the Tertiary (termed epochs) are (oldest to youngest): Paleocene; Eocene, which began about 58 million years ago; Oligocene, which began about 36 million years ago; Miocene, which began about 25 million years ago; and Pliocene, which began less than 10 million years ago. The Quartenary has only two epochs, Pleistocene (about 1.8 million years ago to 10,000 years ago) and Recent (or Holocene—about 10,000 years ago until present).

glenoid fossa—See mandibular fossa.

horn—A frontal appendage with a permanent bony core covered with compressed hairlike material (keratin); the covering is not shed in bovids but is shed annually in antilocaprids.

hypsodont—Pertaining to a high-crowned tooth; such teeth have shallow roots.

incisiform—Having the general form of an incisor.

incisive foramen—(also referred to as anterior palatine foramen or palatine slit). See Fig. 46.

incisor—One of four basic kinds of mammalian teeth that, when present, originate in premaxilla of cranium; anteriormost teeth (those in front of canines) in lower jaw. See Figs. 45 and 47.

inflated—Enlarged or expanded, not flattened or compressed.

infraorbital canal—Canal through zygomatic process of maxilla from wall of orbit to side of rostrum, where it passes through infraorbital foramen. See Fig. 42.

infraorbital foramen—Opening in maxilla from orbit onto the face (rostrum of cranium). See Fig. 45.

interorbital breadth (or interorbital constriction)—See Fig. 47.

interparietal—Unpaired bone of cranium at juncture of paired parietals with supraoccipital. See Fig. 46.

jugal (or malar)—Midbone in zygomatic arch. See Fig. 46.

jugo-maxillary suture—Juncture between jugal and maxillary bones at anterior margin of zygomatic arch.

jugular foramen—See posterior lacerate foramen.

labial—Pertaining to lips; for example labial side of tooth is that side nearer lips rather than tongue; lateral surface of a tooth.

lacrimal—Paired bone of cranium situated in anterior portion of orbit, between jugal and frontal bones; has small opening(s) for tear (lacrimal) duct. See Figs. 45 and 46.

lacrimal pit—Pit or opening in lacrimal bone containing tear duct.

lambdoidal ridge (or lambdoidal crest)—Bony ridge formed at juncture of occiput and parietal bones. See Figs. 45 and 46.

lingual—Pertaining to tongue; for example, lingual side of tooth is that side nearer tongue rather than lips; also medial surface of a tooth.

lyre-shaped—Shaped as the lyre, a musical instrument.

malar—See jugal.

mandible—Lower jaw, formed by paired dentary bones.

mandibular fossa (or glenoid fossa)—Concavity on ventral surface of zygomatic arm of squamosal with which the dentary articulates. See glenoid fossa on Fig. 47.

mastoid—Paired bone of skull, bordered by squamosal, exoccipital, and tympanic. See Fig. 47 for mastoid (mastoidal) breadth.

mastoid bulla—That part of bullar region covered by mastoid (mastoidal) bone. See Fig. 46.

maxilla (or maxillary)—Paired bone of skull situated behind premaxilla and anterior to palatine; this bone bears all upper teeth except incisors. See Figs. 45 and 46.

maxillary toothrow—Length of upper row of teeth, in maxillary bone (thus always excluding incisors); measurement usually taken at alveoli.

mesopterygoid fossa—Shallow area posterior to internal nares and between pterygoid bones.

milk teeth—See deciduous dentition.

Miocene—See geologic time.

molar—One of four basic kinds of mammalian teeth located posterior to premolars; molars have no deciduous precursor. See Figs. 45 and 47.

molariform—Having the general form of a molar; most often used to describe molarlike premolars.

nares—Openings, external and internal, of the nasal passage. See anterior nares on Fig. 47 and posterior nares on Fig. 48.

nasal—Paired bone of cranium situated on anterodorsal surface of skull. See Fig. 46.

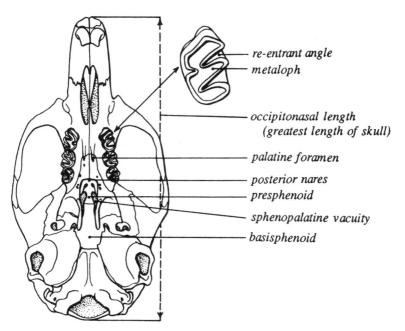

Fig. 48.—Drawing (with some features labeled and one measurement shown) of ventral view, with enlarged first molar, of cranium of *Neotoma* (modified after Hall, 1955).

occipital condyle—Surface of articulation between cranium and first cervical vertebra (atlas); two such condyles in mammals, on either side of foramen magnum. See exoccipital condyle on Fig. 46.

occlusal—Of, or pertaining to, the chewing or shearing surface of a tooth.

orbit—Space (socket for the eye). See Fig. 45.

palatal bridge—Bony tooth-bearing plate (somewhat raised, part of paired maxillae) behind diastema; especially well developed in lagomorphs and rodents.

palatal emargination—Indentation at anterior margin of rostrum (in premaxillae), particularly in bats.

palate—Bony plate in roof of mouth formed by fusion of premaxillae, maxillae, and palatines. See Fig. 43 for measurement.

palatine slit—See incisive foramen.

Paleocene—See geologic time.

palmated—Form like the palm of a hand; term applied to antlers in which at least some spaces between tines are filled with bony growth.

parietal—Paired bone of skull that is situated posterior to frontal and dorsal to squamosal. See Figs. 45 and 46.

paroccipital process—Bony projection extending ventrally from, or located ventrally on, parocciptal bone. See Fig. 46.

postcanine teeth—Collectively, teeth behind canines (premolars and molars).

posterior lacerate foramen (or jugular foramen)—Opening in basicranial region between tympanic bulla and basioccipital.

posterior palatine foramen—Small opening in hard palate near juncture of maxillae and palatines. See palatine foramen on Fig. 48.

postorbital bar—Complete dorsoventral connection (bar) of bone posterior to orbit; results from fusion of postorbital process of frontal and postorbital process of jugal.

postorbital breadth (or postorbital constriction)—See Fig. 47.

postorbital process of frontal—Bony projection of frontal bone posterior to orbit. See Figs. 45 and 47.

postorbital process of jugal—Bony projection of jugal bone posterior to orbit. See Figs. 45 and 47.

preglenoid crest—Bony ridge or shelf on anterior part of glenoid (mandibular) fossa, especially well developed in carnivores.

premaxilla (or premaxillary)—Paired bone in anterior of cranium; point of origin of upper incisors (when present). See Figs. 45 and 46.

premolar—One of four basic kinds of mammalian teeth, situated between canines and molars. See Figs. 45 and 47.

prismatic—Cheekteeth (especially of rodents) with well-developed triangles (or prisms) of enamel surrounding basins of dentine.

process—Small bony projection.

procumbent—Protruding or projecting forward, such as procumbent incisors.

pterygoid—Paired bone on ventral surface of cranium, posterior to palatine and anterior to alisphenoid; forms border of internal nares. See Figs. 45 and 46.

quadrant—One-fourth of the total complement of teeth; one side of upper or one side of lower jaw.

re-entrant angle—Infolds of enamel on the side, front, or posterior part of a tooth. See Fig. 48.

Recent—See geological time.

rostrum—Portion of cranium anterior to orbit. See Fig. 47.

ruminant—Ungulate with specialized four-chambered digestive system; cud-chewing mammals.

sagittal crest—Raised bony ridge on middorsal aspect of cranium; especially well developed in carnivores. See Figs. 45 and 46.

secodont—Tooth that is compressed to increase shearing action; especially well developed in carnassials of some carnivores.

secondary cusp—See cusplet.

selenodont—Cusp pattern of molars in which individual cusps are cresent shaped; highly developed in ungulates.

semiprismatic—Cheekteeth (especially of rodents) with partially closed (slight re-entrant angles) triangles (or prisms) of enamel surrounding dentine.

septum—A partition.

skull—Cranium plus mandible.

squamosal—Paired bone of cranium on posterolateral surface of skull. See Fig. 46.

supraorbital process—Bony projection above orbit on frontal bone; especially well developed in lagomorphs.

supraorbital shelf—Small bony ridge on dorsal margin of orbit on frontal and parietal bones.

suture—Point of contact (or juncture) and fusion between adjacent bones.

temporal ridge—Bony ridge on frontal and parietal bones. See Fig. 46.

truncated—Abruptly or sharply marked, having a square or broad end; sometimes appearing as cut off.

tympanum—Of, or pertaining to, bony ring, as in shrews, that does not form a complete bulla.

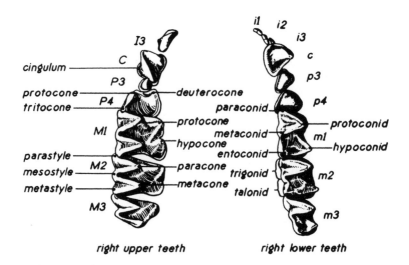

Fig. 49.—Drawings of right upper and lower toothrows of *Tadarida*, with teeth and other features labeled (after Hall, 1955).

tympanic bulla—See auditory bulla.

ungulates—Hoofed mammals, such as artiodactyls.

unicuspid—With single well-developed cusp.

uropatagium—Portion of flight membrane of bats situated between legs; encloses tail or part of tail.

vomer—Unpaired bone of cranium that may form septum in nasal passage.

zygomatic arch (or zygoma)—Arch of bone protecting orbit; formed by jugal and compliments of maxillary and squamosal bones. See Fig. 46.

zygomatic breadth—See Fig. 47.

zygomatic plate—Platelike extension of maxilla in anterior part of zygoma. See Fig. 46.

LITERATURE CITED

Anderson, S., and J. K. Jones, Jr. (eds.). 1984. Orders and families of Recent mammals of the world. John Wiley & Sons, New York, xii + 686 pp.

Bekoff, M. 1977. Canis latrans. Mamm. Species, 79:1-9

Carleton, M. D., and G. G. Musser. 1984. Muroid rodents. Pp. 289-379, *in* Orders and families of Recent mammals of the world (S. Anderson and J. K. Jones, Jr., eds.), John Wiley & Sons, New York, xii + 686 pp.

DeBlase, A. F., and R. E. Martin. 1980. A manual of mammalogy with keys to families of the world. Wm. C. Brown Co., Dubuque, Iowa, xii + 436 pp. [dated 1981 but first distributed in December 1980].

Hall, E. R. 1955. Handbook of mammals of Kansas. Misc. Publ. Mus. Nat. Hist., Univ. Kansas, 7:1-303.

———. 1981. The mammals of North America. John Wiley & Sons, New York, 1:xv + 1-600 +*90* and 2:vi + 601-1181 + *90*.

Jones, J. K., Jr., D. M. Armstrong, R. S. Hoffmann, and C. Jones. 1983. Mammals of the northern Great Plains. Univ. Nebraska Press, Lincoln, xii + 379 pp.

Jones, J. K., Jr., R. S. Hoffmann, D. W. Rice, C. Jones, R. J. Baker, and M. D. Engstrom. 1992. Revised checklist of North American mammals north of Mexico, 1991. Occas. Papers Mus., Texas Tech Univ., 146:1-23.

Klingener, D. 1984. Gliroid and dipodoid rodents. Pp. 381-388, *in* Orders and families of Recent mammals of the world (S. Anderson and J. K. Jones, Jr., eds.), John Wiley & Sons, New York, xii + 686 pp.

Krutzsch, P. H. 1954. North American jumping mice (genus Zapus). Univ. Kansas Publ., Mus. Nat. Hist., 7:349-472.

Marshall, L. G., J. A. Case, and M. O. Woodburne. 1990. Phylogenic relationships of the families of marsupials. Pp. 433-505, *in* Current mammalogy (H. H. Genoways, ed.), Plenum Press, New York, 2:xvii + 1-577.

Nowak, R. M. 1991. Walker's mammals of the world. John Hopkins Univ. Press, Baltimore, 5th ed., 1:xlv + 1-642 + xlix-lxiii and 2:x + 643-1629.

Ryan, M. R. 1989. Comparative myology and phylogenetic systematics of the Heteromyidae. Misc. Publ. Mus. Zool., Univ. Michigan, 176:iv + 1-103.

Vaughan, T. A. 1986. Mammalogy. Saunders Coll. Publishing, Philadelphia, 3rd ed., vii + 576 pp.